Issues in Teaching and Learning of Education for Sustainability

In a fast-changing, globalising world, the teaching and implementation of a curriculum for Education for Sustainability (EfS) has been a challenge for many teachers. *Issues in Teaching and Learning of Education for Sustainability* highlights the issues and challenges educators and academics face in implementing EfS and gives examples of what an EfS curriculum may look like and how some institutions translate the theory into practice.

Organised into three parts, the volume looks at: the who (EfS for whom), the what (EfS curriculum) and the how (translating from theory to practice). The concluding chapter provides ideas and directions on where the world can proceed regarding sustainability education and how it can help in the teaching and learning of sustainability. Considering social issues such as poverty, education, health, culture and the use of natural resources, this book proposes a different path towards Education for Sustainability.

Providing concrete data on the realisation of sustainable development, *Issues in Teaching and Learning of Education for Sustainability* will be of interest to geographers, geography educators and professionals concerned with Education for Sustainability.

Chew-Hung Chang is an Associate Professor at the Humanities and Social Studies Education Academic Group at NIE, NTU. He is a geography educator serving as the Co-Chair of the International Geographical Union – Commission on Geographical Education (IGU-CGE), Co-Editor of the journal *International Research in Geographical and Environmental Education* (IRGEE), as well as the President of the Southeast Asian Geography Association (SEAGA). In addition to being a teacher educator, he has published extensively across areas in geography, climate change education, environmental and sustainability education.

Gillian Kidman is an Associate Professor of Science Education at Monash University, Australia. Her teaching and curriculum design are award-winning and she was a Lead Writer and Senior Advisor for Australia's National Curriculum – Australian Curriculum: Science (Science Inquiry Skills strand). She is the Co-Editor of the journal *International Research in Geographical and Environmental Education* (IRGEE) and is the Australian Representative – Steering Committee, International Geographical Union-Commission Geographical Education (IGU-CGE) and Group Leader – Sub-Committee for Diversifying Research at IGU-CGE. Gillian has research and teaching interests in the sciences and humanities, with a particular interest in inquiry forms of teaching and learning as well as the potential inquiry pedagogies have for the integration of science with other disciplines.

Andy Wi is a research associate at the National Institute of Education, Nanyang Technological University, Singapore and the Assistant Secretary of the Southeast Asian Geography Association (SEAGA). Previously, he was an educator with more than 15 years of teaching experience. He has a PhD in public education, environmental education and policy. His current research focuses are in public and environmental education, geography curriculum and Education for Sustainability. He has published in research journals such as the *Environmental Education Research* (EER) and the *International Research in Geographical and Environmental Education* (IRGEE).

Routledge Research in Education

This series aims to present the latest research from right across the field of education. It is not confined to any particular area or school of thought and seeks to provide coverage of a broad range of topics, theories and issues from around the world.

Recent titles in the series include:

Arts-Based Teaching and Learning in the Literacy Classroom
Cultivating a Critical Aesthetic Practice
Jessica Whitelaw

Quantitative Measures of Mathematical Knowledge
Researching Instruments and Perspectives
Edited by Jonathan Bostic, Erin Krupa, and Jeffrey Shih

Assessment in Mathematics Education Contexts
Theoretical Frameworks and New Directions
Edited by Jonathan Bostic, Erin Krupa, and Jeffrey Shih

Examining the Use of Online Social Networks by Korean Graduate Students
Navigating Intercultural Academic Experiences
Joong-Hwan Oh

Stories from Inequity to Justice in Literacy Education
Confronting Digital Divides
Ernest Morrell and Jennifer Rowsell

Issues in Teaching and Learning of Education for Sustainability
Theory into Practice
Edited by Chew-Hung Chang, Gillian Kidman, and Andy Wi

For a complete list of titles in this series, please visit www.routledge.com/Routledge-Research-in-Education/book-series/SE0393

Issues in Teaching and Learning of Education for Sustainability

Theory into Practice

Edited by Chew-Hung Chang,
Gillian Kidman and Andy Wi

Routledge
Taylor & Francis Group
LONDON AND NEW YORK

First published 2020
by Routledge
2 Park Square, Milton Park, Abingdon, Oxon, OX14 4RN

and by Routledge
52 Vanderbilt Avenue, New York, NY 10017

Routledge is an imprint of the Taylor & Francis Group, an informa business

First issued in paperback 2021

British Library Cataloguing-in-Publication Data
A catalogue record for this book is available from the British Library

Library of Congress Cataloging-in-Publication Data
A catalogue record has been requested for this book

ISBN: 978-1-138-32535-7 (hbk)
ISBN: 978-1-03-208929-4 (pbk)
ISBN: 978-0-429-45043-3 (ebk)

Typeset in Sabon
by Wearset Ltd, Boldon, Tyne and Wear

Contents

Illustrations

Figures

Tables

Contributors

Maria Remedios Abalahin is a general education kindergarten teacher at a network charter school in New York City. Prior to moving to the United States, she was a research assistant at the National Institute of Education Singapore. Her research interests include analysis and evaluation of policies in school leadership, system-wide education reforms and early childhood education policies. Following her true passion for working with young children in their formation and development as conscientious and empowered human beings, she has recently moved to the United States to continue her journey as an educator.

Kaira Zoe Alburo-Cañete specialises in Critical Development Studies focusing on the intersections of gender, environment and disasters. For the past eight years, Kaira has been involved in policy and development research on various contemporary issues and their interplays in the Philippines such as understanding local perspectives of resilience, promoting sexual and reproductive health in humanitarian settings, examining human trafficking following disasters, and supporting gender and disability inclusion in disaster risk management. Kaira is currently undertaking her PhD at the University of New South Wales in Sydney, Australia. Her dissertation examines women's experiences and practices of disaster recovery in post-Haiyan Tacloban City, Philippines.

Yembuu Batchuluun is a Professor in the Department of Geography at Mongolian National University of Education. Her research interests are in geographical education, curriculum, development, innovative teaching and learning, and climate change education, as well as physical geography. She has published widely in these areas. She is founder and an executive director of the Mongolian Association for Geographic Education (MAGE) and Editor-in-Chief of *Geoforum-Mongolia* journal. Professor Batchuluun is also an active participant from Mongolia at the IGU Commission on Geographical Education.

Nina Brendel is an Assistant Professor of Geography Education at the University of Potsdam, Germany. She studied Secondary Education for the subjects Geography and German at the University of Bayreuth (Germany) and at Macquarie University Sydney (Australia). After completing two years of teacher training, she worked as a research assistant at the Institute of Geography Education at the University of Muenster (Germany), where she received her PhD in 2016. Since 2017, she has been working as a 'junior professor' for geographical education at the University of Potsdam (Germany). Her research mainly focuses on Education for Sustainable Development, reflective thinking, digital learning environments and participatory research.

Niranjan Casinader is Senior Lecturer in the Faculty of Education at Monash University, specialising in curriculum, pedagogy and Humanities education through the lenses of culture and problem solving. A geographer by training, he worked in school education for over 30 years before entering academia. In a corollary to his focus there, his research and publications are centred on the relationships between education, culture, thinking, development and globalisation, both contemporary and historical. Current projects include studies into the transcultural capacity and the inquiry literacy of teachers in schools, building upon his model of Cultural Dispositions of Thinking. He has been professionally recognised for his educational work, both locally and internationally.

Chew-Hung Chang is a geography educator serving as the Co-Chair of the International Geographical Union, Commission on Geographical Education, Co-Editor of the journal *International Research in Geographical and Environmental Education*, as well as the President of the Southeast Asian Geography Association. In addition to being a teacher educator, he has published extensively across areas in geography, climate change education, environmental and sustainability education.

William Infante received his Bachelor and Masters degree in Electronics Engineering from Ateneo de Manila University, Philippines. Currently, he is a PhD candidate at the University of Sydney, Australia. He also participated in student exchange programmes at Kyushu University, Nanyang Technological University and the Technical University of Munich. He is a research affiliate at the Ateneo Innovation Center that aims to translate research into sustainable systems, services and engagements through a multidisciplinary approach. His research interests include power system economics, electromobility, and renewable energy integration.

Md Saidul Islam is Associate Professor of Sociology and a Coordinator of the Environment and Sustainability Research Cluster in the School of Social Sciences and Asian School of the Environment, Nanyang Technological University, Singapore (NTU). His research interests include international development and environmental sociology, and he has published five books and over three dozen articles and book chapters on these topics. He is the recipient of a number of awards including Early Investigator Award 2015 (Canadian Sociological Association) and the Outstanding Scientist Award for Publication Excellence 2017 (i-Proclaim Annual Research Award, Malaysia).

Gillian Kidman is Associate Professor of Science Education at Monash University, Australia. Her teaching and curriculum design is award-winning and she was a Lead Writer and Senior Advisor for Australia's National Curriculum – Australian Curriculum: Science (Science Inquiry Skills strand). She is the Co-Editor of the journal *International Research in Geographical and Environmental Education* (IRGEE) and is the Australian Representative – Steering Committee, International Geographical Union-Commission Geographical Education (IGU-CGE) and Group Leader – Sub-Committee for Diversifying Research, (IGU-CGE). Gillian has research and teaching interests in the sciences and humanities, with a particular interest in inquiry forms of teaching and learning as well as the potential inquiry pedagogies have for the integration of Science with other disciplines.

Fa Likitswat is an Instructor of Landscape Architecture Program at Faculty of Architecture and Planning, Thammasat University. Fa has been researching, writing and lecturing on sustainable and ecological landscape related issues from the Winter of 2012. She has focused on education for sustainability in courses including Sustainable Landscape Studio, Landscape Ecology, and Life and Sustainability. Fa holds a Bachelor of Interior Architecture degree from Thammasat University in Thailand and a Master of Landscape Architecture degree from Illinois Institute of Technology (IIT), Chicago. Her project 'The Gravity of Transformation' has won Schiff Foundation Fellowship in 2012.

Jin Ma received his Bachelor and Masters degree in Electrical Engineering from Zhejiang University, Hangzhou, China, a PhD degree in Electrical Engineering from Tsinghua University, Beijing, China, in 1997, 2000 and 2004. He was a faculty member of North China Electric Power University, China from 2004 to 2013. He is currently an Associate Professor at the School of Electrical and Information Engineering, The University of Sydney, Australia. His major research interests are in power system modelling, nonlinear control system, dynamic power system and power system economics. He is a registered Chartered Engineer in the UK and a Member of IEEE and IET.

Liberty Pascua is a PhD student at the University of Sydney in Australia. Prior to her pursuit of higher degree studies, Ms Pascua was a research associate at the National Institute of Education, Singapore. Her research interests include the analysis and evaluation of policies and their implementation in the areas of Education for Sustainable Development, climate change and disaster risk reduction. Ms Pascua has published extensively in academic journals such as the *Journal of Geography*, *Journal of Environmental Education, Evaluation and Program Planning* and the *International Research in Geographical and Environmental Education.*

Nguyen Minh Quang is a Lecturer at School of Education – Can Tho University and Managing Director of Mekong Environment Forum. His research interests include contemporary Vietnam issues such as politics, education reforms, hydropolitics and environmental security. A scholar-practitioner of interdisciplinary studies, he is keen to combine his intellectual capacity with his desire to serve for the sustainable development of the Lower Mekong Region. He is also contributing to The Diplomat and some Vietnam-based independent environment magazines.

Getsel Uranchimeg is a Senior Lecturer of geography at the Mongolian National University of Education, specialising in geographical education, school geography and environmental education. Her current research focuses on the climate change education. As a former teacher of school geography for 20 years, she is also involved in research activities related to teaching standards, teacher learning, assessment of geography and textbook development. She is also co-author of numerous school geography textbooks and publications on geography education.

Andy Wi is a Research Associate at the National Institute of Education, Nanyang Technological University, Singapore and the Assistant Secretary of the Southeast Asian Geography Association. He has a PhD on public education, environmental education and policy. His current research focuses on public and environmental education, geography curriculum and Education for Sustainability.

Qi Zhang is an Associate Professor in the School of Geographic Sciences, East China Normal University, and the vice-chief editor of *Geography Teaching* which is a journal for Geography teachers and researchers. She has been a Geography Education, Environmental Education and Education for Sustainable Development (EE and ESD) researcher and practitioner for more than 20 years. She works as a coordinator and consultant for several international cooperative teachers' training programmes on EE and ESD.

Preface

School curricula have been evolving in keeping abreast with the issues that affect humankind in a fast-changing world. Key elements of changes include increased focus on topics such as environmental change and globalisation. Furthermore, there is a more explicit articulation about the curriculum for Education for Sustainability (EfS) and methods to interpret and implement it in the classrooms. From the EfS conference 2017 in Singapore, this has been a challenge for most teachers, and many have asked if there are examples that they can learn from. This book builds on the previous book 'Education and sustainability: Paradigms, policies and practices in Asia' by the EfS community. The focus is to highlight the issues and challenges educators and academics face in implementing EfS and examples on how some institutes translate it into practice. This book provides a collection of critical pieces that examine what an EfS curriculum may look like and how to translate it from theory to practice. It also answers the important question of EfS; for whom and from whom. Furthermore, this edited book volume begins with a discussion and examines definitions on what Education for Sustainability (EfS) is about. The following parts will be organised to describe the three broad categories which are; for who (EfS for whom?), the what (EfS curriculum) and the how (translating from theory to practice). Part I examines what EfS is and whom it applies to and how the various authors have used it in their countries. Part II discusses what an EfS curriculum looks like and gives examples and strategies on how to develop a sustainability curriculum. Part III provides examples of how countries translate 'sustainability strategies' from theory to practice in the local, global and glocal scales. The concluding chapter provides ideas and directions on where the world can proceed regarding sustainability education and how it can help in the teaching and learning of sustainability.

Abbreviations

ABET	Accreditation Board for Engineering and Technology
ACARA	Australian Curriculum Assessment and Reporting Authority
ASEAN	Association of Southeast Asian Nations
AUN-QA	ASEAN University Network-Quality Assurance
CC-DRR	Child-Centered Disaster Risk Reduction
CCE	Climate Change Education
CDC	Curriculum Development Council
COP21	Conference of Parties 21
DESD	Decade of Education for Sustainable Development
DRR	Disaster Risk Reduction
DRRM	Disaster Risk Reduction Management
EDB	Hong Kong Education and Manpower Bureau
EE	Environmental Education
EfS	Education for Sustainability
ESD	Education for Sustainable Development
FPT	Vietnam's largest information technology service company
HEIs	Higher Education Institutions
HFA	Hyogo Framework for Action's
IDNDP	the International Decade for Natural Disaster Prevention
IGU	International Geographical Union
ITP	International Training Programme
IYGU	International Year of Global Understanding
MOET	Vietnamese Ministry of Education and Training
OECD	Organisation for Economic Co-operation and Development
P3D Mapping	Participatory Community 3-Dimensional Mapping
PDRRMO	Provincial Disaster Risk Reduction Management Office
PISA	Programme for International Student Assessment
PRC	People's Republic of China
RMIT	Royal Melbourne Institute of Technology
S3CDRRM	Strengthening Capacities for Child-Centered Disaster Risk Reduction and Management

SEAMEO	Southeast Asian Ministers of Education Organization
SFDRR	Sendai Framework for Disaster Reduction
UN	United Nations
UNCRC	United Nation Convention on the Rights of the Child.
UNDP	United Nations Development Programme
UNESCO	United Nations Education, Scientific and Cultural Organisation
UNICEF	United Nations International Children's Emergency Fund

1 Defining Education for Sustainability (EfS)

A theoretical framework

Gillian Kidman, Chew-Hung Chang and Andy Wi

Introduction

Over the last decade, the United Nations Education, Scientific and Cultural Organisation (UNESCO) has been promoting Education for Sustainability (EfS) and Education for Sustainable Development (ESD) initiatives to countries at all levels of education, engaging learners with innovative content and pedagogy around themes of sustainable development (Wals, 2014, UNESCO 2012). The growing interest in sustainability has led to an increase in the movement of EfS and ESD (Lozano, Lukman, Lozano, Huisingh, & Lambrechts, 2013). In 2002, the 57th United Nations General Assembly put in place a UN Decade of Education for Sustainable Development, lasting from 2005 to 2014 (UNESCO, 2002). While governments, non-governmental organisations and the academic community have discoursed about the need for EfS, it would seem that there is no consensus on the definition of EfS. Similarly, the definition of ESD has also been highly debated (Chang, 2014). The definitions for both EfS and ESD mean different things to different people under different contexts and situations (Higgitt, 2006).

Merrill (2018) states that the term 'sustainability' refers to goals that promote environmental, economic and social well-being for current and future generations. The Cambridge dictionary (https://dictionary.cambridge.org/dictionary) states that sustainability is 'the quality of causing little or no damage to the environment and therefore able to continue for a long time'. In environmental science, sustainability is defined as 'the quality of not being harmful to the environment or depleting natural resources, and thereby supporting long-term ecological balance' (Dictionary.com).

While there is a common understanding of the term 'sustainability', there is a lack of consensus on the concept of sustainable development, resulting in a wide variety of definitions and interpretations. This is especially true for the terms EfS and ESD in terms of pedagogy and educational paradigms (Merrill, 2018; Sterling 2003, 2004, 2010). In spite of the multiple interpretations, the commonly used definition from the document

'Our Common Future', also known as the Brundtland Report, defines sustainable development as 'development that meets the needs of the present without compromising the ability of future generations to meet their own needs' (Brundtland, 1987; Mebratu, 1998). While we recognise that sustainability and sustainable development refer to different things, we will refer to sustainability as a larger concept including sustainable development, loosely defined.

Similarly, there are also many different interpretations and definitions to EfS. It is not the purpose or intention of the authors to discuss the basis of EfS at length, but to highlight the issues faced by teachers/educators in teaching, and by students in learning about EfS. Before discussing EfS, we must first understand the ontology of EfS definitions.

The Cloud Institute defines EfS as:

> a transformative learning process that equips students, teachers, schools, and informal educators with the knowledge and ways of thinking that society needs to achieve economic prosperity and responsible citizenship while restoring the health of the living systems upon which our lives depend.
>
> (Cloud, 2004, 2010)

A quick search on the internet yielded around 30 different interpretations, philosophies and definitions from different organisations around the world regarding EfS. A word cloud (see Figure 1.1) identifies key themes across the different definitions. From the word cloud, we can see that 'sustainable', 'education', 'needs', 'learning', and 'future' are most commonly used.

As previously indicated, sustainability is often defined as meeting the human needs of the present and ensuring that future generations can also meet their own needs (Brundtland, 1987). This basic definition has been the foundation for many sustainability efforts that have developed around the world. From topics ranging from climate change management, disaster risk reduction, as well as cultural and economic development, sustainability encompasses a wide range of issues. While we know sustainability is important to ensure a safe and possible future generation, the question then is how do we educate people about sustainability? Sustainability cannot be achieved merely by technological solutions, political regulation or financial instruments, but requires a change in the individual's mindset, lifestyle and action. It is believed that there should be quality education and learning for sustainable development at all levels and in all social contexts. Tan and Chang (2008) explain that education is an important process to pass on as a means of sharing knowledge, best practices and to enculturate epistemologies about what constitutes sustainable practices to our future generation. It is an educational approach that aims to develop individuals with the values inherent in sustainable development into all

Figure 1.1 Key themes/words in defining EfS.

aspects of learning, and to encourage changes in behaviour that allow for a more sustainable and just society for all (AESA, 2016; DESD, 2005). Over the years, the United Nations has consistently emphasised the use of education as a tool for advocating sustainable development:

> Education is critical for promoting sustainable development and improving the capacity of people to address environment and development issues.
>
> (UNESCO, 1992, para. 36.3, p. 2)

> Education is critical for achieving environmental and ethical awareness, values and attitudes, skills and behaviour consistent with sustainable development and for effective public participation in decision making. Both formal and non-formal education are indispensable to changing people's attitudes so that they have the capacity to assess and address their sustainable development concerns.
>
> (Agenda 21 – Rio, p. 5)

EfS is the process by which we educate individuals about the values, opportunities and choices each of us has to develop as an informed, independent, responsible and active agent of change in an effort to contribute to the future of our society and ecological systems (Sterling, 2010). It is a transformative learning process that engages and equips students, teachers, educators and learners with the knowledge, attitude, skills and values necessary to contribute and safeguard environmental, social and economic well-being, both in the present and for future generations. Engaging learners to think critically and creatively about the future can achieve this, and are the changes that are needed to improve quality of life across the globe. Similar to our treatment of sustainable development and sustainability, we will refer, for convenience, to EfS as a broader concept, including ESD, loosely defined.

Sterling (2010) posits that the nature of sustainability requires a fundamental change of epistemology and, therefore, education has to meet the accompanying challenges. This implies that changes in curricula, pedagogy, policy and institutional structures are necessary in order to achieve sustainability. While all academics, educators, researchers and policymakers agree that EfS is important and should be included in the curriculum, the question is how do we seek to structure and align our teaching practices with the behavioural and learning needs of students?

There is little guidance in the research literature on a basic framework for the implementation of effective EfS. We attempt to remedy this in the book by bringing together an overview of the issues in teaching and learning of EfS. We included contributions from teachers, educators and researchers working in a variety of education sectors and synthesise their perspectives and experiences about what is EfS, who it is for, the curriculum and how to translate theory into practice. It is through this synthesis that we offer an EfS theoretical framework for teaching and learning. However, before considering this framework, we need to consider why such a framework is needed.

For whom and by whom: who is it for?

In 2016, the International Year of Global Understanding (IYGU) was established to connect culturally different local actions and global challenges to create global sustainability (IYGU, 2016). An initiative of the International Geographical Union (IGU), one of the key messages in IYGU is that 'Everyday actions matter for global climate change'. In other words, it is important to examine how to change individual and social practices with respect to sustainability, as well as redesigning environmental policies through sound science (Werlen, 2015). This means that people need to understand what their daily actions mean for the world as a whole to overcome global challenges (Chang & Wi, 2018). This knowledge may be of relevance to:

- The person on the street;
- A teacher designing a lesson plan;

- A child interested in skills of the future;
- A school leader who needs to rally the teachers;
- Education policy makers (e.g. curriculum designers, ministers), and
- Academic communities in Higher Education Institutions (HEIs) wanting to advance the discourse of sustainability.

Moreover, we need to clarify what we mean about positionality. Several questions arise:

- Is there a realistic way to think about education?
- Is there a pragmatist, realist, neoliberal way of thinking?
- Does it empower our children to form opinions rather than to rote-learn facts (Dewey, 2008)?
- Are we preparing students for the future tomorrow and not today? Is there more than one way of looking at this?

Despite these varying interpretations of EfS, as defined by whom and for whom, there needs to be a working definition that is pragmatic and is able to cut across theoretical paradigms. As noted above we have constructed a theoretical framework, based upon existing theories in the literature, as well as upon the research published within the chapters of this book. The existing literature was taken as a focal point, upon which we interpret our contextual studies. We base the discussions in this book on this framework. Therefore, the examples in this book focus on EfS teaching and learning, and how education can impact education and policy makers towards effective EfS.

Education for sustainability – theoretical framework

EfS not only provides scientific and technical skills, but also the motivation and social support needed for individuals wanting to pursue and apply their scientific and technical skills. For this reason, we need to improve the quality and coverage of environmental education and re-orient its goals to recognise the importance of sustainable development. Many international, regional and national groups have recognised that environmental education is the root of EfS, and have developed their own strategic or action plans. A common call in many international, regional and national environmental education initiatives is the need for all government agencies to collaborate and have a shared understanding of and commitment to policies, strategies and programmes of Education for Sustainable Development.

EfS goes beyond providing information about the environment. It is a process that motivates and engages people in creating sustainable futures. Figure 1.2 illustrates this process. Figure 1.2 is an EfS theoretical framework. It is a blueprint for guiding effective EfS. Each of the authors in this

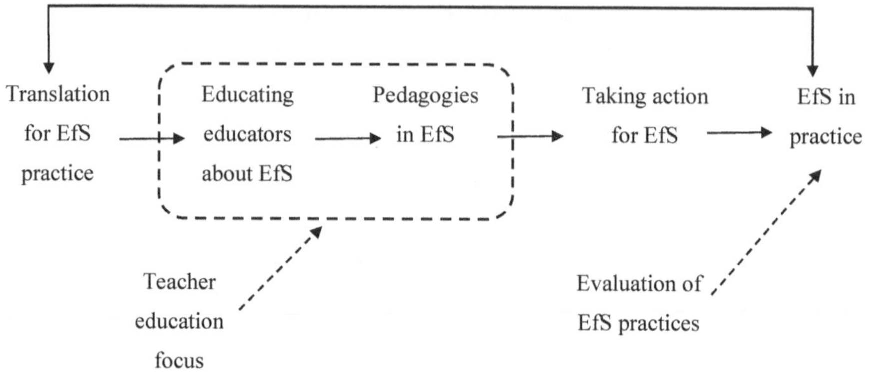

Figure 1.2 The EfS theoretical framework.

book has grounded their work in this framework. Their individual contexts, tools and analyses contribute to a common worldview or lens from which to support our collective thinking and understandings about effective EfS.

Phases within the framework

The EfS theoretical framework involves five phases. The framework is cyclical, in that it loops back on itself when we consider EfS practices. Phases 2 and 3 relate to teacher education, and evaluation of EfS progress ideally occurs in Phase 5. Phases 2 and 3 relate closely to EfS in terms of formal schooling from primary institutions through to tertiary institutions. Educators in Phases 2 and 3 need to understand their work is housed in a larger setting – that of the real world. Educators need to see that their planning and actions need to encourage planning and actions in their students in the real world – an understanding that empowers the learner to act upon their knowledge. As we describe each phase separately below, we consider the confluence of the research literature (the essential ideas found in the reported literature) and we consider the uniqueness of the ideas contained in this book (the new knowledge our authors are adding to the EfS debate).

1 Phase 1 is the *Translation for EfS practice*.
  In this phase, the literature confluence considers:
  - Curriculum integration;
  - Interaction between people and government;
  - The changes that are needed, and
  - The 'drivers' that are deemed as essential.

The unique contributions to this confluence (see Chapters 2, 3, 4, 5, 7, 8, 9, 11, 12, 13) are:

- The specific location benefits of EfS;
- The learning directions of EfS, and
- The empowerment of the individual.

2 Phase 2 is the *Educating educators for EfS.*

In this phase, the literature confluence considers:

- Core learning is needed;
- Public awareness is paramount, and
- Active-learning is essential.

The unique contributions to this confluence (see Chapters 1, 2, 4, 5, 9, 10, 12) are:

- Teaching people to question their actions;
- EfS needs to be location specific, and
- Consideration of both the physical and built environment are essential.

3 Phase 3 relates to the *Pedagogies for EfS.*

In this phase, the literature confluence considers:

- The need for student-centred interactions;
- The need for the delivery of knowledge and values, and
- The quality of experience is critical (attitudinal shift and knowledge balance).

The unique contributions to this confluence (see Chapters 1, 2, 4, 7, 8, 9, 10, 12, 14) are:

- That dynamic solutions are obtainable;
- That action plans are essential;
- That skill differentiation is needed as well as is developed, and
- That spatial skills need to be developed.

4 Phase 4 relates to *Taking action for EfS.*

In this phase, the literature confluence considers:

- Curriculum development;
- Interaction with the environment is essential, and
- Developmental goals are essential.

The unique contributions to this confluence (see Chapters 1, 3, 4, 7, 8, 9, 10, 12, 13) are:

- That personal attitudes need considering, and
- Personal growth in terms of action must be fostered.

5 Phase 5 relates to *EfS in practice.*

In this phase, the literature confluence considers:

- The need for curriculum reforms;
- That the values of the people and government need to align;
- That education, both formal and informal, is needed to enact sustainable practices, and
- That cultural relevance is paramount.

The unique contributions to this confluence (see Chapters 1, 2, 3, 4, 5, 6, 8, 10, 12, 13, 14) are:

- The need for the development of key competencies;
- The demonstration of an ethical approach;
- Of leading by example, and
- An outward-looking approach.

Consequently, this framing of EfS maps out the key issues in enacting EfS and provides a canvas against which the discourses of examples are explored throughout each chapter.

Developing agency among educators

The processes illustrated in Figure 1.2 emphasise the importance of developing agency among educators. Each educator is a key agent for educating and bringing about changes in students' learning behaviour in formal educational settings. However, education is no longer confined to the classroom as formal education but also through online education and informal education. In addition, EfS and ESD encompasses a wide range of learning activities in K-12 education, technical and vocational training and tertiary education, both in formal, non-formal and informal settings. Educators in all settings, both formal and informal, must empower all learners to take action.

Examples of curricula documents

Throughout this book we illustrate such learning activities that explore how we are empowering learners and the wider community to take action for a sustainable future.

Primary

- The importance of fieldwork as a vehicle for sustainability education (Chapter 8);
- The need to create a culture of sustainability and making knowledge available to young children (Chapter 2).

Secondary

- An example of a partnership with the United Nations Children's Fund to implement an initiative entitled 'Strengthening Capacities for Child-Centered Disaster Risk Reduction and Management' (Chapter 3).

Higher education institutions

- How Vietnamese Ministry of Education and Training (MOET) launched a series of education reforms in terms of governance systems, curriculum, mission focus, external relations, research, and financing (Chapter 11);
- The need for a cross-boundary subject (using techniques from one field and borrowing from other fields) and realignment of topics by providing learners with some familiarity for tackling problems in sustainability (Chapter 9);
- A model to diagnose and foster reflective thinking about sustainability in grade 10 and 11 students (Chapter 10);
- The importance of fieldwork as a vehicle for sustainability education (Chapter 8);
- An example of Curriculum Development Stages and Structure on Climate Change Adaptation for pre-service teacher training (Chapter 7);
- The importance of incorporating sustainable learning as well as active-learning for the bigger classes by using a built-environment module as an example (Chapter 4);
- Examining how theory affects students' environmental sustainable practices during Chinese New Year (Chapter 13).

Public education

- A description of how governments should integrate the learning of disaster risk reduction (DRR) in every country's education system (Chapter 6);
- An example of a good climate change education policy is to have interaction between the government and the people (Chapter 5);
- The importance of international cooperation and learning, and a school-based interdisciplinary curriculum on climate change (Chapter 12).

What does a curriculum for EfS look like?

While we see EfS and ESD definitions transcend social, economic, environmental, cultural and political contexts, the implications of sustainable development remain to be seen. However, one key concept that cannot be missed is the interaction between the environment and society; assuming that society encapsulates the social, environment, culture and politics. As previously discussed, the EfS definition should be education that meets these goals for sustainability. Notwithstanding that sustainability is in itself a contested definition from concepts such as sustainable development, EfS should cut across the domains of formal and informal education and within formal education (across levels of primary, secondary and tertiary education).

Clearly, EfS resides on the notion of sustainability. In its root form, sustainability is about sustaining and has the notion of cutting across timescales – whether it is sustaining over a period of five years, a generation or a generation of children to come. Education sociologist Michael Young (Young, Lambert, Roberts, & Roberts, 2014) suggested that there is an interesting relationship between knowledge, curriculum and the future school, or a 'Three Futures' approach to curriculum, which is described broadly by the school and non-school boundaries of school knowledge:

> **Future 1:** Knowledge is inherited and there is little room for change or development. The transmission of knowledge is crucial to maintaining the boundaries of school knowledge in the F1 model.
>
> **Future 2:** The goals of education are to ensure the employability of students when they graduate from schools. With increased focus on applied subjects and vocational training, the school subject disciplines are weakened.
>
> **Future 3:** A middle ground that helps learners engage with the information they encounter, within the contextual understanding of school knowledge, and asks critical questions that will develop deeper understandings of the issues at hand.

As described by Young et al. (2014), disciplinary subjects are supported and challenged by the discoveries of academics and the research community. Disciplinary subjects are supported by teacher knowledge of how different children learn and what are effective activities that will encourage the learners to take their learning further. Therefore, in terms of ESD and EfS, teachers and educators should aim to help individuals develop environmental and ethical awareness, values and attitudes, skills and behaviour consistent with sustainable development for the future benefit of our society.

Parts of this book

The contributors to this book are participants in an international group of higher education practitioners, advocates, academics and educators that have been gathering in Asia since 2013. This community of practice grew out of a multidisciplinary social network, which includes individuals from HEIs and Institutes or Schools of Education, research institutes, government agencies, non-governmental organisations and consultancies around the world.

The research presented in this book draws upon both the formal and informal educational contexts. The research supports the theoretical framework presented in Figure 1.2. In this current chapter, we explored what Education for Sustainability (EfS) is and presented a theoretical framework for EfS. The remaining chapters are organised to present a tripartite progression of sustainability education teaching and learning.

Part I – Chapters 1 to 5. EfS for whom?

Part I considers the importance of identifying the various EfS and ESD stakeholders and helping them understand the concept of sustainability. It is critical to demonstrate that their daily actions can help overcome global challenges. In addition, the chapters in this part also discuss the implementation of sustainability education and action in several countries.

Part II – Chapters 6 to 10. What does an EfS curriculum look like?

Part II explores the design of an EfS curriculum and discusses strategies on how to develop a sustainability curriculum. It also addresses the opportunities and challenges of developing an EfS curriculum, educating educators for EfS and pedagogies for EfS.

Part III – Chapters 11 to 14. From theory to practice (translating)

This part includes descriptions of implementing EfS in both the local and global setting in terms of translating theory to practice. It provides ideas and directions on how the world can proceed in terms of sustainability education and how it can help in the teaching and learning of sustainability.

Summary

EfS is still in its infancy. While it is successful in particular projects and sectors, its application beyond the UN Decade of Education for Sustainable Development will need a more strategic and coordinated framework guiding the use of education and learning, to embed systemic change within the community, institutions, government and industry.

Environmental education for sustainability incorporates a perspective on looking to the future. This future dimension further distinguishes EfS from conventional approaches to environmental education. Central to this approach is an examination of probable futures resulting from existing environmental relationships and of possible alternative environmental futures. Crucially linked to this future dimension is the concept of 'empowerment', and 'action'. Empowerment and action are crucial factors in the

EfS theoretical framework offered in this chapter (see Figure 1.2) and are explored in subsequent chapters of this book.

References

Australian Education for Sustainability Alliance (AESA). (2016). Getting Started with Sustainability in Schools: What is Education for Sustainability? Accessed from www.sustainabilityinschools.edu.au/what-is-efs.

Brundtland. (1987). Report of the World Commission on Environment and Development: 'Our Common Future': United Nations. Accessed from www.un-documents.net/our-common-future.pdf.

Chang, C. H. (2014). *Climate Change Education: Knowing, Doing and Being*. Abingdon: Routledge.

Chang, C. H., & Wi, A. (2018). Why the World Needs Geography Knowledge in Global Understanding: An Evaluation from a Climate Change Perspective. In Demirci, A., González, R. M., & Bednarz, S. W. (Eds.), *Geography Education for Global Understanding* (pp. 29–42). Cham: Springer.

Cloud, J. P. (2004). Education for sustainability: What is its core content? *NAAEE Communicator, 1*(10), n.p.

Cloud, J. P. (2010). *Educating for a Sustainable Future*. Curriculum 21: Essential Education for a Changing World, Chapter 10. Washington, DC: ASCD.

Decade of Education for Sustainable Development (DESD). (2005). United Nations Decade of Education for Sustainable Development 2005–2014. Retrieved from http://unesdoc.unesco.org/images/0013/001399/139937e.pdf.

Dewey, J. (2008). *Experience and Education*. New York: Simon & Schuster.

Higgitt, D. (2006). Finding space for education for sustainable development in the enterprise economy. *Journal of Geography in Higher Education, 30*(2), 251–262.

IYGU. (2016). IYGU Key Messages. Retrieved 19 May 2017 from www.global-understanding.info/what-is-iygu/iygu-key-messages/.

Lozano, R., Lukman, R., Lozano, F. J., Huisingh, D., & Lambrechts, W. (2013). Declarations for sustainability in higher education: Becoming better leaders, through addressing the university system. *Journal of Cleaner Production, 48*, 10–19.

Mebratu, D. (1998). Sustainability and sustainable development: Historical and conceptual review. *Environmental Impact Assessment Review, 18*(6), 493–520.

Merrill, M. Y. (2018). Education for Sustainability in Asian Context. In Merrill, M. Y., Burkhardt-Holm, P., Chang, C. H., Islam, M. S., & Chang, Y. (Eds.), *Education and Sustainability: Paradigms, Policies and Practices in Asia*. Abingdon: Routledge.

Sterling, S. (2003). Whole systems thinking as a basis for paradigm change in education: Explorations in the context of sustainability. PhD, University of Bath, UK.

Sterling, S. (2004). An analysis of the development of sustainability education internationally: Evolution, interpretation and transformative potential. In *The Sustainability Curriculum: The Challenge for Higher Education* (pp. 43–62). London: Routledge.

Sterling, S. (2010). Learning for resilience, or the resilient learner? Towards a necessary reconciliation in a paradigm of sustainable education. *Environmental Education Research, 16*(5–6), 511–528.

Tan, I. G. C., & Chang, C. H. (2008). Geography education for sustainable development in Southeast Asia. *International Research in Geographical and Environmental Education, 17*(4), 289–291.

United Nations Educational, Scientific and Cultural Organization (UNESCO). (1992). UN Conference on Environment and Development: Agenda 21 (Switzerland, UNESCO).

United Nations Educational, Scientific and Cultural Organization (UNESCO). (2002). UN Decade of Education for Sustainable Development. Retrieved from https://en.unesco.org/themes/education-sustainable-development/what-is-esd/un-decade-of-esd.

United Nations Educational, Scientific and Cultural Organization (UNESCO). (2012). Shaping the Education of Tomorrow: 2012 Report on the UN Decade of Education for Sustainable Development, Abridged. 7, place de Fontenoy, 75352 Paris 07 SP, France.

United Nations University (UNU). (2016). Regional Centre of Expertise on Education for Sustainable Development Web Portal. United Nations University. Retrieved 1 March 2016 from www.rce-network.org/portal/home.

Wals, A. E. (2014). Sustainability in higher education in the context of the UN DESD: A review of learning and institutionalization processes. *Journal of Cleaner Production, 62*, 8–15. doi:10.1016/j.jclepro.2013.06.007.

Werlen, B. (2015). International year of global understanding. An interview with Benno Werlen. *Journal of Research and Didactics in Geography, 2*(4), n.p.

Young, M., Lambert, D., Roberts, C., & Roberts, M. (2014). *Knowledge and the Future School: Curriculum and Social Justice*. London: Bloomsbury Publishing.

Part I

EfS for whom?

2 Sustainability in Primary Geography

Maria Remedios Abalahin and Chew-Hung Chang

Introduction

Education for sustainability in Primary Geography

Globalisation has a profound impact on education institutions to meet the demands of twenty-first century skills (Jickling, 2005). To cope with complexities and uncertainties in our globalised world, new forms of learning are emerging and evolving (UNESCO, 2012) and interdisciplinary knowledge becomes important in helping our children make sense of the world. In addition, a diverse range of environmental issues around the world dominates the headlines of everyday news. From its emergence as a formal concept in the 1970s through the Stockholm Declaration of 1972, environmental education (EE) began to shift focus by the early 1990s (Tillbury, 1995). In addition to discourses on knowledge, skills and attitudes that could empower our youth for environmental action, the discourses on Education for Sustainable Development (ESD) began to appear in the 1990s. With the UN declaration of the Decade of ESD (DESD) in 2002, a plurality of approaches to understanding sustainable development has led to equally diverse approaches to Education for Sustainability (EfS). While there are nuanced and arguably substantial differences among the concepts of EE, ESD and EfS at the ontological and epistemological levels, there is in addition, a proliferation of ways of teaching and learning about the human-environment interaction. With the increasing interest in issues of changes to the Earth's climate, another body of discourse on climate change education (CCE) has also emerged. These approaches are evident in the global discourse with the inclusion of sustainable development – e.g. Agenda 21 and Earth Charter in 1992 and DESD. Following this trend, we look into the teaching of EE, CCE and EfS, subsumed under the *interdisciplinarity* of Education for Sustainable Development (ESD) (J. M. Davis et al., 2008; Smith et al., 2015; UNESCO, 2012), to examine the current status of teaching of sustainability in Primary Geography. In order not to create confusion for the reader, the authors will be referring to ESD as the main organising concept for teaching sustainability in Primary Geography.

Increasing capacity and improving ESD, therefore, is of paramount priority set against the backdrop of emerging global discourses on sustainable development. The 2012 UNESCO (UNESCO, 2012) report on the role of education in achieving the SDGs found the following:

- ESD is increasingly perceived as a catalyst for innovation in education.
- Boundaries between schools, universities, communities and the private sector are blurring due to a number of trends. ESD is often at the heart of new, creative multi-stakeholder configurations involving these ESD stakeholders.
- As ESD progresses, a co-evolution of pedagogy is occurring. It appears that as the sustainability content of the curriculum evolves, pedagogy is evolving simultaneously.

Post-2015 development agendas, however, still reflect an 'undervaluing of education as [a] vehicle for social change [and that] most educational programmes do not yet reflect the purposes and goals of sustainable development' (Sterling, 2016, p. 210). Changes in socio-economic structures, dynamics in global economies and the role of education in building a sustainable world has implications for teacher training. Haubrich (2000, p. 284) reckoned that, in addition to playing the roles of 'social workers, moderators, advisors and facilitators', teachers now have a responsibility for educating children for 'complex and diverse' changes in life in the future 'at every scale from local to global'. ESD, beyond building knowledge, endeavours to change mindsets and behaviour (Jucker, 2011; Laurie, Nonoyama-Tarumi, Mckeown, & Hopkins, 2016). In order to develop meaningful and relevant ESD for all, we have to first identify the issues and gaps in policy curriculum and programmatic curriculum for ESD in primary, especially beyond just helping children raise awareness and build knowledge.

Some studies in the literature have reported on children's environmental knowledge, understanding, concerns and values but Catling (2005) argued that this is not enough. For instance, Palmer and Birch (2004) reported that little has been written for teachers of early years on 'the origins of children's subject knowledge and conceptual development in the areas of geography and environmental education' (p. 8). The deliberate focus on children's geography, and in this case Primary Geography, stems from a response to Catling's (2005) compelling argument for listening to children's voices, and raising issues for the development of Primary Geography research.

Chambers (1999) proposed that an important area in the research agenda for environmental education is '... learning how knowledge, values and attitudes can be transformed into action for the environment whilst at the same time avoiding indoctrination' (p. 92). We propose that such is the case for ESD in geography education as well. Concepts such as place-based education (Casey, 1993, 1997/1999), bio-regional nature immersion (Hensley, 2011), eco-literacy (Goleman, Bennet, & Barlow, 2012) and powerful knowledge (Roberts, 2014; Young, 2008) in teaching ESD are foundational to teaching geography. Haubrich (2000) further argued that the field of geography offers the knowledge and discipline necessary to understand emerging realities of a globalised world wherein addressing current issues of sustainable development are of the highest importance.

The aims of ESD are aligned with Dewey's idea that universal education is about teaching children to become critical thinkers and informed decision makers (Caiman & Lundegård, 2013; Dewey, 1997; Jucker, 2011). Primary Geography offers opportunities for teachers to design learning based on disciplinary knowledge, and to ground the interdisciplinarity of knowledge and pedagogies of ESD (Eilam & Trop, 2010) to achieve its expected outcomes (UNESCO, 2012). To scale this, however, from the classroom to national and global context adds another layer of complexity to ESD in Primary Geography. Young and Muller (Young & Muller, 2010) envisioned three futures of education and argued for the importance of recognising the 'differentiatedness' of knowledge in the curriculum (Future 3) as the ideal set-up to successfully scale the impact of ESD on the local level to the national and global level. Indeed, education and ESD in particular is empowering within the Future 3 context as new forms of knowledge are being produced at the boundary between everyday knowledge and disciplinary knowledge. A simple observation that certain countries have lower carbon emissions can be made relevant only in examining if there are additional contributions to carbon emissions from the flow of goods and services globally. Surely importing more goods from another country where carbon emissions are higher will make the country in question look good, notwithstanding the current account balances for the economy in question.

Although national curricula for school geography are often contextualised within the local sphere, current trends in educational policy and reforms indicate that there are undeniable global processes and dynamics at work in the development of national curriculum (Grindsted, 2015; Jickling, 2005; Pauw, 2015; Young & Muller, 2010). Butt and Lambert (2014) illustrated how national education policies and reforms on educational funding and curriculum matters profoundly affect the future of school subjects, making a case for geography education in particular. Similarly, there is a need to understand how ESD situates within geography education curriculum in schools as well as on a national and global scale. This is how geography education can 'contribute to the joint goals of personal happiness and the sustainable development of the world' (Haubrich, 2000, p. 284). There are educational opportunities by examining the scaling of these issues from the classroom to the local and then the global contexts.

Glonacal agency heuristic

Incorporating or integrating global ESD goals onto national policies is evidently an increasing trend in the past decade. Olsson, Gericke, and Chang Rundgren (2016) emphasised that SD needs to be described for each of its dimensions with respect to their interrelation both in time (past–present–future) and in space (near-far). *Glonacal agency heuristics* (Marginson & Rhoades, 2002) is a useful analytical framework to understand these emergent dynamics and how the global rhetoric on ESD has influenced national education reforms. It impacts schools at the local level in aspects such as accountability, teacher training and curriculum implementation (Aikens, McKenzie, & Vaughter, 2016; Bengtsson & Östman, 2014; Buchcic & Grodzińska-Jurczak, 2004; J. M. Davis et al., 2008; Jucker, 2011; Læssøe & Mochizuki, 2015; Laurie et al., 2016). This framework offers a multidimensional lens needed for the 'exploration and analysis of types and patterns of influence and activity, to conceptualize social relations and actions globally, nationally and locally (Marginson & Rhoades, 2002, p. 290)'.

Marginson and Dang (2016) expanded Vygotsky's sociocultural theory of learning and contextualised it within globalisation:

> Vygotsky's thought, applied in the context of global spatiality, provides one useful framework for drawing macro-historical changes and world relations together with changes at the level of individuals in society: in their modes of thought, imagined possibilities, and agency freedoms.

This is what glonacal agency heuristic affords us as we examine the policy and programmatic curriculum of sustainability in Primary Geography – understanding the agency of various social elements (e.g. policy makers, teachers, students) at the intersection of global, national and local dimensions. How has global sustainability influenced national curriculum reforms? How are these reforms translated to teaching practice and classroom dynamics at the implementation phase? Sterling (2016) commented that the critical role of education in enhancing the effectiveness of ESD delivery and integration is 'through developing informed engagement, agency and empowerment among all affected stakeholders, and through unlocking and fostering their creativity, ideas, abilities and enthusiasm' and to 'build lasting change, that is, "sustainable change", because it is owned and affected by participating stakeholders and learners' (p. 11).

Primary Geography as geography for young children

The learning and experience of primary school children cannot escape the impact of globalisation. '[T]he increasing mobility of people, and the rising flood of daily information through a variety of different media such as television, mobile phones, and the Internet' (Schmeinck, 2013, p. 399) typify the lived experiences of our children. We cannot under-estimate the amount of knowledge that children have. Catling (2005) suggests that we should recognise that very young children have 'embryonic geographical knowledge and understanding' (p. 298) that are crucial to developing understandings in ESD. Indeed Palmer and Birch (2004) have reported that children at the age of four and five years old are able to describe:

> their sense of hot and cold places, refer to local experiences and places beyond their experience, and indicate their concerns and values for the environment; do hold misconceptions, and have inaccurate, even false, information, and are limited in their environmental awareness; but a few, can, even at four years old, refer to aspects of environmental processes, such as recycling, from family experience.
>
> (Palmer & Birch, 2004 cited in Catling, 2005, p. 298)

Studies show that the idea of sustainable consumption is the most easily grasped sustainability concept and a valid point of introduction on sustainability for young children (Siraj-Blatchford, Smith, & Samuelson, 2010).

Schmeinck (2013) suggests that primary school children live in 'places, in different spaces, together with different people and in relation to their environment. They experience and engage with and within different situations, with spatial and everyday life connections locally and with distant places' (p. 399). Indeed the agency of the child at the local scale cannot be ignored in discussing issues of ESD in primary schools. The increasingly urban settings within which children grow up in and the rapid development of technology that allows humans to be connected to cyber space and staying interconnected has caused certain disconnect among children with the natural world. Indeed, 'this kind of deteriorating connection to the natural world at the same time as we're accelerating this interconnectedness, to me, is the deep source of a lot of the most, the greatest imbalances that we all live with' (Senge, 2015).

While conventional knowledge of primary education refers to Primary 1 to 6 or grades 1 to 6, this particular chapter argues for K to Primary 3, extending it to early childhood education as a means of understanding Primary Geography as a continuum. In the context of ESD, *The Gothenburg Recommendations on Education for Sustainable Development* positions it for Primary Geography within the continuum of lifespan learning. This identifies early childhood as a 'natural starting point' for all ongoing Education for Sustainability, and educators as playing a significant role during these early years (J. M. Davis et al., 2008).

ESD empowers children to be active participants in their community. It is a transformative education that nurtures children to become problem seekers, problem solvers and action takers in their own environment. A student who can start with his or her 'embryonic geographical knowledge and understandings' (Catling, 2005) early in life and 'think of things as connected, even if he/[she] revises his[/her] view with every succeeding year, has begun the life of learning' (Van Doren, 1965).

Literacy development in young children is therefore critical in empowering and building agency. The earlier children are exposed to the complexities of the world they live in, processed using developmentally-appropriate approaches, the sooner they develop the level of understanding they need to successfully navigate these complexities and make empowered choices towards sustainability. By doing so, we as educators avoid the 'blind spots' that undermine the agency of the child by neglecting to reconstruct anthropocentric worldviews (J. Davis & Elliot, 2009, p. 13).

What does ESD look like for young children?

The emerging rhetoric among scholars is that sustainability is not a school subject on its own but a learning process of reframing how we see the world and imagine the future, embedded in existing or emerging disciplines such as geography or climate change education. This approach may help build our capacity to live more sustainably and meet emergent challenges in an increasingly complex and uncertain world (Clarke, 2012; Eilam & Trop, 2010; Hensley, 2011). Corcoran and Wals (2004) elucidated, '[s]ustainability is not just another issue to be added to an overcrowded curriculum, but a gateway to a different view of curriculum, of pedagogy, of organizational change, of policy and particularly of ethos' (p. 5).

In a sense, the school is a major stakeholder in the glonacal agency heuristic that we are discoursing here. There is a need to re-orient our current education model (Hensley, 2011) to what Clarke suggested as a re-imagination of school of sustainability 'from an established education system that is primarily defined by the values associated with economic productiveness to one that seeks to meaningfully integrate social and community context into learning transactions' (Clarke, 2012, p. 107).

There is mounting evidence that supports the value that ESD adds to the teaching and learning process in schools and how it can 'convert the learning environment from a sterile environment of drudgery to something more dynamic and meaningful' (Hensley, 2011, p. 120). An examination of research carried out in 18 countries in various continents, worth detailing at length below, illustrates that ESD significantly contributes to improving quality of education (Laurie et al., 2016, pp. 234–235).

At the student level, 11 reports (Australia, Belgium, Canada, China, England, Mongolia, the Netherlands, Peru, Scotland, Taiwan and the US) indicated that students find an ESD approach increases the relevance of their learning content. ESD is reported to give more meaning to a school curriculum that is well adapted to local themes and priorities.

ESD also creates a more interesting learning context for students (Australia, Belgium, the Netherlands, Scotland, Sweden, the US) and helps students recognise their roles in society (Korea). In addition, the increased curricular relevance associated with ESD leads to increased student engagement and commitment (Canada, China, Germany, Japan, the Netherlands, Peru, Scotland, Sweden, the US), self-confidence (Germany), self-esteem (Finland, Korea) and self-awareness (Estonia, Korea).

Further, ESD perspectives foster further engagement with ESD topics in students' private lives, including lifelong learning, informal learning and social entrepreneurship (Finland, Germany). ESD promotes students' spiritual development and helps them treat their surrounding world and cultural heritage with sensitivity, care and respect (Latvia). Finally, ESD perspectives also increase student influence on their education (Scotland, Sweden), while projects such as twinning schools in different countries increase empathy, understanding, stewardship and intellectual engagement (Canada), extending the agency of the child beyond the local to the global.

These empirical studies have described how ESD can potentially contribute to the education of young people, but a gnawing question of how this can be done remains unanswered. Teachers who are at the forefront of the classroom process need to be able to operationalise the reframing of the learning process. While many countries' and regions' school curricula are designed along school subject lines, teachers also need some disciplinary, if not interdisciplinary, lenses to make decisions about how to sequence topics, select resources, design activities and even administer assessment. Is there a more concrete approach to using a subject discipline to teach ESD? Is there a disciplinary way of knowing that could help children make sense of what they learn about sustainability? Geography is a pertinent example of this argument because it is so much a part of children's lives from their earliest years, since without their engagement in knowing and understanding the physical and human everyday worlds, not only would they not undertake such apparently straightforward matters as way-finding but they would not construct their sense of their environments as lifeworlds in order to make use of the affordances they offer. Geography is a fundamental and essentially powerful aspect of being human from the earliest years.

In envisioning the three-futures curriculum, Young and Muller (2010) asserted the importance of recognising epistemological constraints and the crucial role of education in ensuring equity of access to powerful knowledge. Pauw (2015) argues, 'Futures-oriented school geography requires the use of powerful knowledge and an awareness of the situatedness of all knowledge. It is also a natural setting for pedagogical innovation.' Indeed, the central argument for the powerful disciplinary knowledge in geography as a way of reframing ESD is in addressing the issue of how such an approach could potentially help our children manage issues they face in a F3 curriculum. It refers to the relationship between knowledge and curriculum that empowers our children to not only survive but to do well in a world where there are uncertainties and complexities in the environmental, social, economic and cultural issues that they face both at the local and global scales. In answering the question of what does ESD look like for young children in each country, we have to examine this relationship between ESD and the curriculum and how this is expressed through school geography, or not.

Education for sustainability in the national curriculum across various countries

At this point, we shall examine the national curriculum across several countries in various regions around the world to get a general picture of the extent of integration of ESD. This is also at the nexus of the local-global scale of analysis that is adopted in this case. We present an analysis of policy documents and position papers using the futures curriculum framework. Countries selected for this part are those that have implemented national curriculum reforms that coincide with or come after the UN's declaration on the Decade of Education for Sustainable Development 2005–2014 (UNESCO, 2005). This period serves as a reference point for the global call to action to include ESD in national policy education discourse. Here we map out how global rhetoric has informed national education policies. We then examine how these changes at the policy level contributed to the national rhetoric on sustainability in various learning areas, with particular attention to Primary Geography that will finally trickle down to the local scale of the everyday experience of the young child.

Australia

ESD is embedded in the Australian National Curriculum through its Cross-Curricular Priority which 'provide dimensions which will enrich the curriculum through development of considered and focused content that fits naturally within learning areas' (ACARA, 2015b). Australian Curriculum Assessment and Reporting Authority (ACARA) identified the three key concepts for sustainability priority: systems, worldviews and futures. The council employed a multisectoral approach to policy curriculum development. They consulted with experts in the field of sustainability. They involved multiple organisations that advocate national programmes for sustainability, youth empowerment on climate change, and environmental educators. Practising teachers from each learning area with specialisation in sustainability were also consulted to provide expert advice from their respective fields (ACARA, 2015a).

Primary Geography in the Australian National Curriculum falls within the Humanities and Social Sciences learning area. Sustainability priority is integrated across the curriculum by using each learning area's knowledge content to develop worldviews and gives students learning opportunities to better understand various influences on human interaction and its impact on the environment (ACARA, 2015b).

Hong Kong

Primary General Studies (GS) curriculum reform in Hong Kong took effect in 2011 (Curriculum Development Council (CDC), 2011). Although sustainability is not explicitly mentioned in the document, concepts strongly linked to sustainability such as 'develop curiosity and interest in the natural and technological world as well as understand the impact of science and technology on society' and 'develop care and concern for the environment' are explicitly stated in the curriculum aims (CDC, 2011, p. 9). Geography is not a separate learning silo nor stated as a separate learning area. According to EDB, the GS:

> is designed in the notion that students' learning experiences should be connected and not compartmentalised, so that students can develop a holistic view of themselves as individuals in the community, their place in the natural world, and the interaction of human beings with the environment.
>
> (EDB, 2013)

Primary Geography curriculum may be under the key learning area (KLA) People, Social, and Humanities Education and ESD is integrated in principle through the aforementioned GS curriculum aims.

In the national-local dimension, Hong Kong implemented policy reform on school leadership to strengthen implementation of the national curriculum on the ground level. In 2003, the Hong Kong Education Bureau posted 'curriculum leaders' to provide support for teachers and school principals during the medium-term phase of the curriculum reform in 2010/2011 (Hong Kong Education and Manpower Bureau (EDB), 2003). An examination of the perception of curriculum leaders on this recent reform reflects a gradual shift in the dynamics on the national-local dimension in curriculum development and implementation from a top-down approach, with the primary school master/mistress (curriculum development), or PSMCD, serving as the conduit for the national education board to the school principals (Cheung & Yuen, 2017). One may infer that the position of the PSMCD offers a more collaborative dynamics at the national-local dimension, thereby increasing teacher agency at the policy curriculum level. It is, however, unclear from these documents the degree of collaboration in the planning and development for Primary Geography. This is a possible research area that the community can further explore.

UK and Europe

In the United Kingdom, the national curriculum places Primary Geography under the generic label 'Programmes of Study' that include citizenship education, art and design, history, music and physical education (United

Kingdom Department of Education (DfE), 2013b). Published in 2013 and revised in 2014, the Geography National Curriculum (GNC) for Key Stage 1 (5–7 years old) and Key Stage 2 (7–11 years old) states the following aims that are arguably concepts of sustainability (United Kingdom Department of Education (DfE), 2013a):

- develop contextual knowledge of the location of globally significant places – both terrestrial and marine – including their defining physical and human characteristics and how these provide a geographical context for understanding the actions of processes
- understand the processes that give rise to key physical and human geographical features of the world, how these are interdependent and how they bring about spatial variation and change over time

This may illustrate an integration of sustainability in the UK's revised GNC.

Singapore

The primary school social studies curriculum in Singapore aims to equip students with the civic competencies that will empower them to become active, responsive, informed and participative citizens (MOE, 2012a). The primary science curriculum aims to develop the student as an inquirer to help them navigate the natural and physical worlds (MOE, 2013a). Although the topic of sustainability was not explicitly taught to students in primary school, they have been introduced in the form of key geographical concepts such as physical processes, environment, scale and interdependence.

By Primary 4, students will have learned about the life cycles of plants in science under the topic 'Cycles in Plants and Animals' (MOE, 2013c). In the science syllabus, primary school students by Primary 6 will have learned to describe that human interaction with the environment can yield positive and negative impacts under the topic of 'Interactions within the Environment' (MOE, 2013d). Students would also have learned the importance of care and concern for the environment.

In the social studies subject, students will learn that they should care for places around them and the environment under the topic of 'Knowing Myself, Others and My Surroundings' in Primary 1 (MOE, 2012b). Students should be able to appreciate and care for the places around them as value-based outcomes. In Primary 3, students will learn about how they can change the physical environment to meet their needs and the consequences of their actions on the environment and some ways that they can protect the environment, how they can protect the environment as individuals and as groups under the topic of 'Understanding Singapore' (MOE, 2012c). The expected outcomes in values are to appreciate the physical environment that they live in and how they can conserve to protect the environment.

Overview of cases

In summarising the cases presented above, the national curricula cited have included the contextual knowledge and the aspirations for an active citizenry, though through varied approaches. However, we refer to the work of Benavot (2014) for an overview of more recent developments in ESD for primary and secondary education curriculum in various countries. In his analysis of polices and national curriculum documents, he argues that there is currently a strong link between policy intent for ESD to be integrated in the national curriculum but fails to ground this intent in the national or local documents. The effect, therefore, is that ESD is relegated as a mere 'add-on' to existing curriculum content as opposed to allowing it to be an integrated strategy to improve quality learning outcomes (pp. 6–7).

The examples above illustrate how national geography curriculum incorporates sustainability, either as a learning lens or as foundational concept, but the integration at the programmatic level is still unclear. The inherent complexity of the scope, dimension, and spheres of ESD results in a rhetoric-pedagogy gap (Eilam & Trop, 2010) which complicates even further when integrated with fields of knowledge such as geography. A plethora of issues arises in teaching sustainability, presenting various challenges to effectively integrate ESD in existing curriculum frameworks for Primary Geography Education. For example, teachers' epistemological belief or understanding of environmental education, and ESD by extension, affects their teaching practices (Cheng & So, 2014). Issues such as pre-service training (Kennelly, Taylor, & Maxwell, 2008), teacher capacity (Bertschy, Künzli, & Lehmann, 2013; Catling & Martin, 2011) educational tools and instructional methods (Buchcic & Grodzińska-Jurczak, 2004). In Spain, for instance, concepts related to sustainability are vastly present in the teacher training curricula, albeit in a diffused and superficial manner (Sureda-Negre, Oliver-Trobat, Catalan-Fernández, & Comas-Forgas, 2014). There is also the issue of assessment of sustainability learning in Primary Geography. Nevertheless, these challenges persist through different curricula but affect the actual practice differently as curricula are developed by different people from different social, cultural, professional and political contexts.

How are these curricula developed? Who gets to decide?

We return to glonacal agency heuristics (Marginson & Rhoades, 2002) to understand the patterns of influence and interactions of agencies, both as human institutions and as a human activity in integrating ESD in Primary Geography. The reality is that global discourse demonstrates significant impact on policy development at the national level but the trickle-down effect to the local level is far more complicated. Policy curriculum does not necessarily translate to better schools and improved classroom practices that advance the cause of ESD (Læssøe & Mochizuki, 2015; Laurie et al., 2016; Olsson et al., 2016). In most cases, teachers find education reforms to be cumbersome, considering all the additional work and compliance measures on top of all their teaching and administrative duties (Blum, 2008).

Beyond the intersecting *glonacal dimension* in developing policy and programmatic curriculum for ESD, integrating it into Primary Geography must also take into consideration the *glonacal spheres* within which these dynamics happen (Marginson & Rhoades, 2002). Kemmis and Mutton (2012) referred to it as 'practice architectures' of ESD. Green and Somerville (2015) argued that support for these practice architectures increase the chances of successfully integrating sustainability in Primary Geography; teachers will remain reluctant if existing practices do not change.

Agency and capacity for teaching ESD in Primary Geography

Understanding and empowering children's agency is critical to harnessing the transformative power of ESD (Caiman & Lundegård, 2013; Engdahl, 2015; Ideland & Malmberg, 2014; Kadji-Beltran, Zachariou, & Stevenson, 2013; Samuelsson & Kaga, 2008). The earlier ESD as a learning framework to develop critical thinking is introduced to young learners, the greater and more enduring the impact it can have (Davies et al., 2009). Catling (2014, p. 332), illustrating the interplay between teacher agency and child agency, argued for children as 'contributors to our shared knowledge and understanding of the world'; to evaluate classroom dynamics so that primary school children are not treated as 'recipients' or 'beneficiaries' of 'hand-me-down' curricula which emerge from bodies of ideas designed for secondary schooling and are then diluted until suitable for primary consumption.

Davies and her colleagues proposed an action plan for teaching ESD for the inclusion of young children as active participants to help 'ensure the relevance of content to children's everyday lives and their development as active citizens of sustainability' (Davies et al., 2009). When children are guided well through appropriate ESD pedagogies and encouraged to be 'problem seekers, problem solvers, and action takers in their own environment' they become 'significant players in the changes needed for creating sustainable futures' (J. Davis & Elliot, 2009, p. 12).

The local dimension of the *glonacal* agency heuristics frames the dynamics between powerful knowledge and pedagogical choices (Catling & Martin, 2011), consequently affecting teacher agency. As stated in the CGE 2016, 'Geography' helps people to think critically about sustainable living locally and globally and how to act accordingly. Geography is much more than learning many facets and concepts. Its focus is on the patterns and processes that help us understand an ever-changing planet (International Geographical Union, 2016, pp. 10–11).

Eilam and Trop (2010) observed that although there has been significant progress in policy discourse and curriculum development for ESD, continued implementation of old pedagogies derails its advancement. Corney and Reid (2007) argued that the value-laden nature of ESD requires an 'interactive' approach to teaching, and a key consideration for teacher training is the teachers' personal choice of stance in teaching ESD. Pedagogical choices are informed by teacher histories, professional learning as well as their level of engagement in sustainability issues (Corney & Reid, 2007). There is, therefore, a need to reconnect academic knowledge and everyday knowledge of ESD (Catling & Martin, 2011) when bridging the theory-practice gap in teaching ESD in Primary Geography.

Furthermore, teacher agency is a critical component to ESD in Primary Geography. Teachers' agency in the curriculum making process refers to support for meaningful selection of content and material, design of activities and even ways of assessing learning (Chang & Wi, 2018). Close collaboration of teachers with other stakeholders in the community is important in the process of developing curriculum for ESD (Corney & Reid, 2007).

Data from several countries indicate that there is a need to increase teacher capacity through institutional support (Ferreira, Ryan, & Tillbury, 2006; Nelson, 2010; Summers, Corney, & Childs, 2003; Sureda-Negre et al., 2014). In some cases where there is a lack of state-mandated regulatory tools, 'norm supporting structures' (Læssøe & Mochizuki, 2015) and changes in 'practice architectures' (Kemmis & Mutton, 2012) become imperative to help address issues and narrow practice gaps in teaching sustainability in Primary Geography. Indeed, a child-oriented perspective – the capacity to communicate and process with children the complexities of ESD at the level and content that is accessible to the child – is necessary for shared sustainable thinking (Engdahl, 2015).

Discussion and conclusion

ESD can be described as the enactment of transformative, empowering and participative education around sustainability issues, topics and experiences within the early childhood context (Davis, 2015, p. 22). It is essentially laying the foundations to create a 'culture of sustainability'; establishing the 'agency' of the child in co-creating this culture. With learning as a continuum, this still applies to primary education and beyond.

Sustainability in Primary Geography needs to be explicitly stated in programmatic curriculum (Davies et al., 2009). Policies are in place but sustainability as a 'vehicle idea' and learning process is still lacking in

Primary Geography (Benavot, 2014). Although sustainability is implicit in geography and they are interconnected by the same principles of interdisciplinarity and multidimensionality (studying changes in space over time), attaining sustainability aims of transformative learning, empowerment and participative education needs further examination.

Teacher and child agency also play integral roles in the transformative power of ESD, especially in impacting change through local and global scales. How can they make sense of the ripple effect of their choices and actions for sustainability on the national and global dimensions if they are disengaged at the local dimension? This disengagement may also be the source of professional frustration for geography teachers. Unless the inherent wealth of personal histories and powerful knowledge are tapped and reconciled with strategies for professional development, i.e. teacher agency, building capacity for teaching ESD in Primary Geography at the local and national dimension will continue to fall behind the advances made in the global dimension. What we ideally aim for is for the local and national dimensions to be strong enough to create a ripple effect in the global dimensions.

Some may argue that in order to have successful ESD, we must make knowledge available to children so that they can make informed choices about their lives. In referring to the works of Young (2008), we argue that knowledge is in itself a necessary but insufficient condition to young people to engage with issues critically and to take action for what they believe in. We argue that ESD should not be based on a logical argument that humans are to be blamed for the environmental and social issues of our day and hence we as citizens of the global community should do something about it. There is an ethical discourse that we have not given much attention to. Indeed, what logical reason need there be to convince anyone not to waste food? In an analogy of someone being shot by a poisoned dart, the immediate action should be to seek treatment at the nearest medical facility rather than to find out who shot the dart, if the shooter was male or female, and so on. The immediacy of action for ESD cannot be understated. However, we do have to know what poison is used, if we come back to the analogy of the poison dart. Children must be equipped with knowledge, skills, and dispositions that will empower them to make informed and critical decisions about the information that they come into contact with on issues of sustainability. It is only through solid and strong foundations that they can handle issues that are increasingly complex and unpredictable. Teachers have a strong role to play in this. Before we can develop curricula and instructional activities that will engage our students in meaningful ESD, teachers must themselves be grounded in a deep understanding of the issues. It is only with deep knowledge that teachers can effectively choose resources, represent information, sequence lessons, design activities, implement assessment and review their plans effectively for ESD. With deep knowledge and professionalism, there is hope for teachers to regain their confidence and reclaim their agency in ESD for our young children in primary schools.

References

Aikens, K., McKenzie, M., & Vaughter, P. (2016). Environmental and sustainability education policy research: A systematic review of methodological and thematic trends. *Environmental Education Research*, *22*(3), 333–359. doi:10.1080/13504622.2015.1135418.

Australian Curriculum Assessment and Reporting Authority (ACARA). (2015a). *Draft Changes to Australian Curriculum F-10 Consultation Report*. Australia: ACARA.

Australian Curriculum Assessment and Reporting Authority (ACARA). (2015b). *Tracked Changes to F-10 Australian Curriculum*. Australia: ACARA.

Benavot, A. (2014). *Education for Sustainable Development in Primary and Secondary Education* (pp. 1–43). Paris: UNESCO.

Bengtsson, S. L., & Östman, L. O. (2014). Globalisation and education for sustainable development: Exploring the global in motion. *Environmental Education Research, 22*(1), 1–20. doi:10.1080/13504622.2014.989960.

Bertschy, F., Künzli, C., & Lehmann, M. (2013). Teachers' competencies for the implementation of educational offers in the field of education for sustainable development. *Sustainability, 5*(12), 5067–5080. doi:10.3390/su5125067.

Blum, N. (2008). Environmental education in Costa Rica: Building a framework for sustainable development? *International Journal of Educational Development, 28*(3), 348–358.

Buchcic, E., & Grodzińska-Jurczak, M. (2004). Environmental education in Polish primary schools. *International Research in Geographical and Environmental Education, 13*(3), 264–268. doi:10.1080/10382040408668521.

Butt, G., & Lambert, D. (2014). International perspectives on the future of geography education: An analysis of national curricula and standards. *International Research in Geographical and Environmental Education, 23*(1), 1–12. doi:10.1080/10382046.2013.858402.

Caiman, C., & Lundegård, I. (2013). Pre-school children's agency in learning for sustainable development. *Environmental Education Research, 20*(4), 437–459. doi:10.1080/13504622.2013.812722.

Casey, E. S. (1993). *Getting Back into Place: Toward a Renewed Understanding of the Place-world. Studies in Continental Thought*. Bloomington: Indian University Press.

Casey, E. S. (1997/1999). *The Fate of Place: A Philosophical History. A Centennial Book*. California: University of California Press.

Catling, S. (2005). Children's personal geographies and the English primary school geography curriculum. *Children's Geographies*, *3*(3), 325–344.

Catling, S. (2014). Giving younger children voice in primary geography: Empowering pedagogy – A personal perspective. *International Research in Geographical and Environmental Education, 23*(4), 350–372. doi:10.1080/10382046.2014.946321.

Catling, S., & Martin, F. (2011). Contesting powerful knowledge: The primary geography curriculum as an articulation between academic and children's (ethno-)geographies. *Curriculum Journal, 22*(3), 317–335. doi:10.1080/0958517 6.2011.601624.

Chambers, B. (1999). Environmental education and primary geography. *International Research in Geographical and Environmental Education, 8*(1), 90–93. doi:10.1080/10382049908667597.

Chang, C. H., & Wi, A. (2018). Why the World Needs Geography Knowledge in Global Understanding: An Evaluation from a Climate Change Perspective. In Demirci, A., González, R. M., & Bednarz, S. W. (Eds.), *Geography Education for Global Understanding* (pp. 29–42). Cham: Springer.

Cheng, I. N. Y., & So, W. W. M. (2014). Teachers' environmental literacy and teaching – Stories of three Hong Kong primary school teachers. *International Research in Geographical and Environmental Education, 24*(1), 58–79. doi:10.1 080/10382046.2014.967111.

Cheung, A. C. K., & Yuen, T. W. W. (2017). Examining the perceptions of curriculum leaders on primary school reform. *Educational Management Administration & Leadership, 45*(6), 1020–1039. doi:10.1177/1741143215587303.

Clarke, P. (2012). *Education for Sustainability: Becoming Naturally Smart.* Oxon: Routledge.

Corcoran, P. B., & Wals, A. E. (Eds.). (2004). *Higher Education and the Challenges of Sustainability: Problems, Promise, and Practice.* Dordrecht: Kluwer Academic.

Corney, G., & Reid, A. (2007). Student teachers' learning about subject matter and pedagogy in education for sustainable development. *Environmental Education Research, 13*(1), 33–54. doi:10.1080/13504620601122632.

Curriculum Development Council (CDC). (2011). *General Studies for Primary Schools Curriculum Guide (Primary 1–Primary 6).* Hong Kong: Hong Kong Curriculum Development Council.

Davies, J., Engdahl, I., Otieno, L., Pramling-Samuelson, I., Siraj-Blachtford, J., & Vallabh, P. (2009). Early childhood education for sustainability recommendations for development. *International Journal of Early Childhood, 41*(2), 113–117.

Davis, J. (Ed.). (2015). *Young Children and the Environment* (2nd ed.). Victoria, Australia: Cambridge University Press.

Davis, J., & Elliot, S. (2009). Exploring the resistance: An Australian perspective on educating for sustainability in early childhood. *International Journal of Early Childhood, 41*(2), 65–77.

Davis, J. M., Engdahl, I., Otieno, L., Pramling Samuelsson, I., Siraj-Blatchford, J., & Valladh, P. (2008). *The Gothenburg Recommendations on Education for Sustainable Development.* Gothenburg: Swedish International Centre for Education for Sustainable Development (SWEDESD), Gothenburg University.

Dewey, J. (1997). *Democracy and Education.* Göteborg: Daidalos.

Eilam, E., & Trop, T. (2010). ESD pedagogy: A guide for the perplexed. *The Journal of Environmental Education, 42*(1), 43–64. doi:10.1080/ 00958961003674665.

Engdahl, I. (2015). Early childhood education for sustainability: The OMEP world project. *International Journal of Early Childhood, 47*(3), 347–366. doi:10.1007/ s13158-015-0149-6.

Ferreira, J.-A., Ryan, L., & Tilbury, D. (2006). *Whole-School Approaches to Sustainability: A Review of Models for Professional Development in Pre-service Teacher Education.* North Ryde, NSW: Macquarie University.

Goleman, D., Bennet, L., & Barlow, Z. (2012). *Ecoliterate: How Educators are Cultivating Emotional, Social, and Ecological Intelligence.* San Francisco: Jossey-Bass.

Green, M., & Somerville, M. (2015). Sustainability education: Researching practice in primary schools. *Environmental Education Research, 21*(6), 832–845. doi:10.1080/13504622.2014.923382.

Grindsted, T. S. (2015). The Matter of Geography in Education for Sustainable Development: The Case of Danish University Geography. In *Transformative Approaches to Sustainable Development at Universities* (pp. 13–24). Cham: Springer.

Haubrich, H. (2000). Guest editorial: Sustainable learning in geography for the 21st century. *International Research in Geographical and Environmental Education, 9*(4), 279–284. doi:10.1080/10382040008667660.

Hensley, N. (2011). *Curriculum Studies Gone Wild.* New York: Peter Lang.

Hong Kong Education and Manpower Bureau (EDB). (2003). *Education and Manpower Bureau Circular No. 9/2003: Provision of an Additional Teacher Post for Leading Curriculum Development in Primary Schools for Five Years for the 2003/04 School Year.* Hong Kong: Hong Kong Education and Manpower Bureau.

Hong Kong Education and Manpower Bureau (EDB). (2013). *General Studies for Primary Schools.* 27 June 2017. Retrieved from www.edb.gov.hk/en/curriculum-development/kla/general-studies-for-primary/index.html.

Ideland, M., & Malmberg, C. (2014). Governing 'eco-certified children' through pastoral power: Critical perspectives on education for sustainable development. *Environmental Education Research, 21*(2), 173–182. doi:10.1080/13504622.2013.879696.

International Geographical Union. (2016). *2016 International Charter on Geographical Education.* Amsterdam: Commission on Geographical Education.

Jickling, B. (2005). Sustainable development in a globalizing world: A few cautions. *Policy Futures in Education, 3*(3), 251–259.

Jucker, R. (2011). ESD between systemic change and bureaucratic obfuscation: Some reflections on environmental education and education for sustainable development in Switzerland. *Journal of Education for Sustainable Development, 5*(1), 39–60. doi:10.1177/097340821000500109.

Kadji-Beltran, C., Zachariou, A., & Stevenson, R. B. (2013). Leading sustainable schools: Exploring the role of primary school principals. *Environmental Education Research, 19*(3), 303–323. doi:10.1080/13504622.2012.692770.

Kemmis, S., & Mutton, R. (2012). Education for Sustainability (EfS): Practice and practice architectures. *Environmental Education Research, 18*(2), 187–207. doi:10.1080/13504622.2011.596929.

Kennelly, J., Taylor, N., & Maxwell, T. W. (2008). Addressing the challenge of preparing Australian pre-service primary teachers in environmental education: An evaluation of a dedicated unit. *Journal of Education for Sustainable Development, 2*(2), 141–156.

Læssøe, J., & Mochizuki, Y. (2015). Recent trends in national policy on education for sustainable development and climate change education. *Journal of Education for Sustainable Development, 9*(1), 27–43. doi:10.1177/0973408215569112.

Laurie, R., Nonoyama-Tarumi, Y., Mckeown, R., & Hopkins, C. (2016). Contributions of education for sustainable development (ESD) to quality education: A

synthesis of research. *Journal of Education for Sustainable Development, 10*(2), 226–242.
Marginson, S., & Dang, T. K. A. (2016). Vygotsky's sociocultural theory in the context of globalization. *Asia Pacific Journal of Education*, 1–14. doi:10.1080/02188791.2016.1216827.
Marginson, S., & Rhoades, G. (2002). Beyond national states, markets, and systems of higher education: A glonacal agency heuristic. *Higher Education, 43*(3), 281–309.
Ministry of Education, Singapore (MOE). (2012). *Primary Social Studies Syllabus 2012*. Retrieved from www.moe.gov.sg/docs/default-source/document/education/syllabuses/humanities/files/2012-social-studies-(primary)-syllabus.pdf.
Ministry of Education, Singapore (MOE). (2013). *Science Syllabus (Primary)*. Retrieved from www.moe.gov.sg/docs/default-source/document/education/syllabuses/sciences/files/science-primary-2014.pdf.
Olsson, D., Gericke, N., & Chang Rundgren, S. N. (2016). The effect of implementation of education for sustainable development in Swedish compulsory schools – Assessing pupils' sustainability consciousness. *Environmental Education Research, 22*(2), 176–202. doi:10.1080/13504622.2015.1005057.
Palmer, J., & Birch, J. (2004). *Geography in the Early Years*. London: Routledge-Falmer.
Pauw, I. (2015). Educating for the future: The position of school geography. *International Research in Geographical and Environmental Education, 24*(4), 307–324. doi:10.1080/10382046.2015.1086103.
Roberts, M. (2014). Powerful knowledge and geographical education. *The Curriculum Journal, 25*(2), 187–209. doi:10.1080/09585176.2014.894481.
Samuelsson, I. P., & Kaga, Y. (Eds.). (2008). *The Contribution of Early Childhood Education to a Sustainable Society* (pp. 1–136). Paris: UNESCO.
Schmeinck, D. (2013). 'They are like us' – Teaching about Europe through the eyes of children. *Education 3–13, 41*(4), 398–409.
Senge, P. (2015, 2 October). *Systems Thinking in a Digital World*. Paper presented at the Heart-Mind 2015: Human Connection in a Digital World, Vancouver Convention Center, Vancouver, British Columbia.
Siraj-Blatchford, J., Smith, K. C., & Samuelsson, I. P. (2010). *Education for Sustainable Development in the Early Years*. Gothenburg: World Organization for Early Childhood Education. Retrieved from http://old.worldomep.org/wp-content/uploads/2013/12/combined_ESD_book.pdf.
Smith, K., Adriance, P., Mueller, A., Mckeown, R., Nolet, V., Rowe, D., & Vorva, M. (2015). *The Status of Education for Sustainable Development (ESD) in the United States: A 2015 Report to the U.S. Department of State*. Retrieved from www.sustainabilityprofessionals.org/sites/default/files/ESD%20in%20the%20United%20States%20final.pdf.
Sterling, S. (2016). A commentary on education and sustainable development goals. *Journal of Education for Sustainable Development, 10*(2), 208–213. doi:10.1177/0973408216661886.
Summers, M., Corney, G., & Childs, A. N. N. (2003). Teaching sustainable development in primary schools: An empirical study of issues for teachers. *Environmental Education Research, 9*(3), 327–346. doi:10.1080/13504620303458.
Sureda-Negre, J., Oliver-Trobat, M., Catalan-Fernández, A., & Comas-Forgas, R. (2014). Environmental education for sustainability in the curriculum of primary

teacher training in Spain. *International Research in Geographical and Environmental Education, 23*(4), 281–293. doi:10.1080/10382046.2014.946322.

Tillbury, D. (1995). New focus of environmental education in the 1990s. *Environmental Education Research, 1*(2), 195–213.

United Kingdom Department of Education (DfE). (2013a). *National Curriculum in England: Geography Programmes of Study.* Retrieved from www.gov.uk/government/publications/national-curriculum-in-england-geography-programmes-of-study/national-curriculum-in-england-geography-programmes-of-study.

United Kingdom Department of Education (DfE). (2013b). *Primary Curriculum Key Stage 1.* Retrieved from www.gov.uk/education/primary-curriculum-key-stage-1#/education/primary-curriculum-key-stage-1-programmes-of-study.

United Nations Educational, Scientific and Cultural Organization (UNESCO). (2005). *Guidelines and Recommendations for Reorienting Teacher Education to Address Sustainability. Education for Sustainable Development in Action. Technical Paper No. 2.* Paris: UNESCO.

United Nations Educational, Scientific and Cultural Organization (UNESCO). (2012). *Shaping the Education for Tomorrow: 2012 Full-length Report on the UN Decade of Education for Sustainable Development.* Paris: UNESCO.

Van Doren, M. (1965). *Liberal Education.* Boston, MA: Beacon Press.

Young, M. (2008). *Bringing Knowledge Back In: From Social Constructivism to Social Realism in the Sociology of Education.* London: Routledge.

Young, M., & Muller, J. (2010). Three educational scenarios for the future: Lessons from the sociology of knowledge. *European Journal of Education, 45*(1), 11–27.

3 Learning from Haiyan

Translating children's voices into action for resilience

Kaira Zoe Alburo-Cañete

Introduction

Children are often regarded as 'vulnerable' to disasters. However, while their vulnerabilities to disaster are recognised, their participation in disaster risk management processes is often overlooked since conventional approaches remain paternalistic and disempowering for children (Mitchell, Haynes, Hall, Choong, & Oven, 2008). Often with top-down technocratic approaches that have dominated the field of disaster management practice (Hewitt, 1983), children are rendered invisible, passive and voiceless in making decisions regarding disaster preparedness, prevention and mitigation in their own communities. This exclusionary approach towards children and disasters actually 'threatens their safety when disaster strikes and ignores a valuable resource for risk communication, education, advocacy, and practical risk reduction activities' (Mitchell et al., 2008, p. 255).

With this recognition, a number of disaster risk management practitioners, humanitarian agencies and child rights advocate institutions have carried out initiatives that aim to include children in disaster risk reduction strategising and implementation in order to demonstrate how children can be mobilised to increase the resilience and sustainability of their respective communities. The importance then of documenting and sharing good practices, innovations and lessons learned for developing and implementing child-centred or child-focused approaches to disaster risk reduction is underscored (Back, Cameron, & Tanner, 2009). This chapter aims to contribute to producing knowledge about practices in child-centred disaster risk management as well as generating lessons for relevant practitioners.

In 2014, A2D Project – Research Group for Alternatives to Development Inc., a local non-government organisation based in Cebu, Philippines partnered with the United Nations Children's Fund to implement an initiative entitled 'Strengthening Capacities for Child-Centered Disaster Risk Reduction and Management' or S3CDRRM in two communities that were heavily affected by Typhoon Haiyan in 2013. This chapter presents the results of a documentation project that closely followed the implementation of this year-long initiative. Through field observations, individual interviews, focus group discussions and a review of project documents, this chapter identifies the key achievements, strategies and learnings gained in promoting, institutionalising and sustaining children's participation in disaster risk reduction. By highlighting the strategies and outcomes of implementing a child-centred disaster risk management project, this chapter highlights emerging initiatives, policies and programmes that engender child participation in disaster risk management processes.

In the following sections of this chapter, I provide a brief overview of child-centred disaster risk management, its principles and rationale. Second, I present the contextual background of the project as a response to the destruction caused by Typhoon Haiyan in 2013. I then present the processes, strategies and innovations undertaken by the project that contributes to efforts to promote child-centred disaster risk management as a good practice. Lastly, I conclude with highlighting the key takeaways in carrying out child-centred disaster risk management based on the lessons learned from the S3CDRRM Project.

What is child-centred disaster risk reduction?

According to the United Nations Convention on the Rights of the Child, 'a child means every human being below the age of 18 years unless under the law applicable to the child, majority is attained earlier' (United Nations, 1990, p. 1). With the recognition of disasters as located at the intersection of hazards and underlying physical and social vulnerabilities of populations considered at risk (Wisner, Cannon, Davis, & Blaikie, 2004), disaster risk management can be viewed as a development concern and requires us to address the physical, dependency-related, and social vulnerabilities (Peek, 2008) of children as well as upholding their rights as enshrined in the UN Convention on the Rights of the Child (UNCRC). The development of child-centred disaster risk reduction is anchored on the belief that

> [a]lthough children are at special risk in disasters, they are not passive victims ... children and youth can participate in disaster preparedness activities in their homes, schools, and communities that would likely minimise some of the risks that they face.
>
> (Peek, 2008, p. 4)

Hence, with the growing recognition of children's rights and agency in disaster risk reduction, child-centred disaster risk reduction (CC-DRR) is increasingly being promoted and integrated into current programmes and approaches.

At the heart of CC-DRR are four basic principles found in the UNCRC (United Nations, 1990): (1) non-discrimination; (2) best interest of the child; (3) the right to life, survival and development; and (4) the view of the child. In adherence to these, a number of CC-DRR initiatives have demonstrated the benefits to children. As highlighted by Tanner, Garcia, Lazcano, Tribunalo, and Seballos (2009), children are able to conceptualise and analyse risks that affect their lives and communities. When given the chance to take part in risk assessments and analysis, children can mobilise themselves to be agents of change in their own communities and design as well as implement their own projects. Specifically, where children have been involved in disaster risk reduction activities, they have proven to be effective communicators of risk and risk management options. In connection with this, children's groups have been able to mobilise people (and resources) in their communities for disaster risk reduction and leveraged their skills in building social networks and capital (ibid.).

In sum, CC-DRR can be understood as a

> flexible rights-based approach combining child-focused (for children) and child-led (by children) activities with interventions geared towards bringing about change in community, local and national duty bearers. It applies strategies such as awareness raising, capacity-building, group formation, institutional development, research and influencing and advocacy across a range of arenas.
>
> (Plan UK, 2010, p. 3)

The S3CDRRM Project: context and approach

The S3CDRRM Project is an initiative developed as a response to the destructive effects of Typhoon Haiyan (locally known as Yolanda) which tore through Central Philippines in November 2013. Having been considered as the strongest typhoon to have made landfall in recorded history at the time, Haiyan left in its wake more than 6,000 fatalities, over 1 million homes destroyed, around 4 million people displaced, 16 million people affected, with around 5 million of these being children. Throughout the disaster response and recovery phase, emphasis on building community resilience resonated.

The island province of Cebu was among the areas badly hit by Typhoon Haiyan. According to the Cebu Provincial Disaster Risk Reduction Management Office (PDRRMO), the most seriously affected municipalities were located in the Northern parts of Cebu Province. An estimated 1,680,000 people had been affected in 36 municipalities. Among those extremely affected were the municipalities of Daanbantayan and Bantayan. As coastal municipalities which were directly hit by the typhoon, strong winds, heavy rain and storm surges destroyed more than 90 per cent of infrastructure, and displaced thousands. The S3CDRRM Project, in close collaboration with UNICEF, implemented the project in two barangays (villages) in barangay Poblacion in Daanbantayan and barangay Patao in Bantayan.

While there may be various strategies to undertaking child-centred resilience as demonstrated in a number of related case studies, the project adopted a whole-of-community approach to building capacities towards disaster preparedness while at the same time being mindful of engendering spaces for children's engagement and participation in DRR processes. In this regard, the project aimed to maximise children's abilities to creatively communicate risk reduction messages and influence their respective families, schools and communities in the process.

The S3CDRRM Project operated at different levels of engagement to ensure that the initiative had adequate institutional support. To ensure that children's roles in DRRM are recognised, partnership with municipal and barangay local government units, schools, teachers and parents was critical to carrying out the overall project design. Through a participatory, multi-level and child-sensitive approach, capacities for implementing CC-DRRM were developed and strengthened (see Figure 3.1 below).

With the belief that children are key actors in transforming communities towards disaster resilience, the project adopted the following strategies to promote CC-DRRM in partner communities:

- Developing capacity of local governments in CC-DRRM
  The project carried out a series of capacity development training activities for municipal and barangay local government highlighting the need to involve children in DRR processes. The training aimed to ensure that DRR is prioritised at various levels of governance, that plans are developed to form the basis for implementation and that consequentially CC-DRRM is promoted through the development of a conductive operational environment.

Figure 3.1 The S3CDRRM approach.

- Increasing community knowledge of disaster risks
  With children playing a central role, the project facilitated assessments of disaster risks to enable communities to undertake measures for disaster prevention, mitigation and preparedness. Project results in this area highlight the effectiveness of child-led initiatives in conducting school safety audits, community mapping and modelling, and risk assessment.
- Raised public awareness on disaster preparedness and building a culture of resilence
  S3CDRRM sought to make use of risk information, innovation and education to promote a culture of safety in partner communities. By maximising children's abilities as risk communicators, educational materials using various media were developed by the youth. Children played a key role in conducting community campaigns through mobile film showing, performance and visual arts. Social media was also developed as a platform to promote CC-DRRM.
- Nurturing the capacity of children to become leaders in disaster prevention and preparedness
  Through S3CDRRM, student groups and out-of-school youth organisations were formed as the main vehicles for undertaking child-led activities. The project also conducted a number of communication skills. DRR youth leaders also underwent training such as Emergency First Aid and Water Search and Rescue. A youth camp was held to develop young people's abilities to actively participate in governance and become resilience leaders in their own communities.
- Facilitating youth-led children's dialogues with local government officials and school administration
  Youth groups organised through S3CDRRM facilitated the passage of municipal ordinances establishing an award system for barangays with good practices in CC-DRRM. This initiative envisions the continued promotion of a child-centred approach to risk reduction and development. Resulting from the children's dialogues, DRR action plans for schools were formulated.
- Ensuring institutional support for CC-DRRM
  To sustain the gains of the project, S3CDRRM facilitated the passage of municipal ordinances establishing an award system for barangays with good practices in CC-DRRM. This initiative envisions the continued promotion of a child-centred approach to risk reduction in the partner municipalities even beyond the project life. Moreover, Youth-managed DRR resource centres were established with the support of barangay local governments to ensure access to knowledge about disasters, risks and good practices in DRR as well as provide an avenue for children and youth to meet and discuss issues that concern them and their communities.

Working with children, producing results

Given the success achieved by S3CDRRM in terms of achieving its objectives, the project has generated a number of good practices which have contributed to the widespread and rapid uptake of CC-DRR initiatives in partner communities. However, this chapter will focus on good practices that feature notable innovations that contribute to the knowledge and practice of CC-DRR. Essential to these good practices has been the commitment and dedication of children and youth, with the support of their families, teachers and local officials, in taking on the challenge of resilience-building in the aftermath of Typhoon Haiyan.

Good practice 1 – from local to global: raising children's voices for climate action

The destruction brought by Typhoon Haiyan had undoubtedly heightened the call for urgent climate action both locally and globally. As part of the project's mobilisation strategy, S3CDRRM initiated a campaign for participation in the Earth Hour on 28 March 2015. Earth Hour is a global campaign initiated by World Wildlife Fund wherein for one hour every year, households and companies from all over the globe are encouraged to turn off their lights to signify their support for environmental protection and climate action.

As part of the build-up activities for the event, youth leaders launched a social media photo campaign with children and youth calling for a binding treaty for the Conference of Parties 21 (COP21) in Paris. Through the internet, young people who survived the force of Haiyan called on world leaders to take concrete actions to address climate change. With the support of local governments in the two partner municipalities, the Earth Hour event was participated in by over 1,500 children, youth, other community members and local officials. With the massive support gained for carrying out the Earth Hour Campaign, the event served to raise community awareness regarding their experiences as Haiyan survivors and existing concerns over climate change as well as emphasising the need for action at local, national and international levels. With the visible participation and leadership of children during the event, partner communities, including schools and local barangay governments, recognised that children are capable and have important roles in contributing to risk reduction.

The following were the lessons learned from the activities:

- Greater appreciation, understanding and participation in issues of scale (e.g. disasters and climate change) is achieved by connecting global concerns to local experiences.
- When children are able to demonstrate their capacities, leadership and creativity in advancing campaigns such as the Earth Hour, it helps

transform previously paternalistic and disempowering views regarding children's agency.

- Information technology is an effective platform to popularise and share initiatives on resilience-building and the role of children.

Good practice 2 – disaster risk reduction through schools: an entry point for community mobilisation

The project initially involved two schools in the partner communities, namely Patao National High School and Daanbantayan National High School. Activities conducted in and through both schools included:

- Organising student associations called Earth Savers' Clubs; conducting school safety audits;
- Peer learning through room-to-room campaigns led by Earth Savers' Club members;
- Carrying out risk communication through songs, visual and performing arts;
- Conducting community preparedness drills (with schools playing an important role);
- Linking with out-of-school youth, and
- Undertaking youth-led initiatives such as coastal clean-ups and other disaster prevention/mitigation activities.

It has long been recognised that schools offer great opportunities to promote disaster risk reduction and resilience in the wider community. Thus, DRRM is advanced not only in schools, but schools in general are a vehicle to reach out and influence community institutions and stakeholders to prioritise risk reduction. This is due in part to the role of schools in bridging knowledge and practical local action by providing space for children and youth to discuss concepts of hazards and risks, transmitting risk information, and physically protecting communities from disasters (Campbell & Yates, 2006).

With the success of youth mobilisation in schools to promote greater community resilience, the project has seen a massive uptake in a further nine schools in Bantayan and Daanbantayan, namely Doong National High School, St Paul's Academy, Baod Elementary School, San Jose Elementary School, Malingin National High School, Tapilon National High School, Paypay National High School, SK Luis Canete National High School and Tominjao National High School.

The following were the lessons learned from the activities:

- Schools provide a good entry point to promoting resilience in the communities due mainly to the resources available for knowledge sharing as well as the strategic role they play in communities.

- Mobilising youth for peer education both in-school and out-of-school has been instrumental in promoting children's roles in DRRM.
- S3CDRRM's strategy of mobilising students and coordinating with and getting the support of teachers and school administration strengthened the role of schools as an entry point for raising community awareness on disaster and climate risks.

Good practice 3 – children's dialogues: creating spaces for children's participation in disaster risk reduction and development

A series of children's dialogues were facilitated through the S3CDRRM Project to engender spaces for children's participation in issues that concern them, especially in relation to disaster risk reduction and development in general. These dialogues were led by children mobilised through the project and participated in by school administration personnel, local officials, representatives of humanitarian organisations, emergency service providers and other key stakeholders in DRRM. Apart from discussions on DRRM concepts and issues, these dialogues also included exhibits featuring children's visual artworks (i.e. posters) on their perspectives on various environmental issues that they consider important. Children have also served as resource speakers to a number of these assemblies.

The children's dialogues became an important platform for children to engage with community institutions and local governments and advance their agenda for resilience-building. Among the most significant results of children's engagements with these stakeholders was the institutionalisation of youth representation in barangay and municipal DRRM councils in partner communities in Bantayan and Daanbantayan. Moreover, the continued children's active participation in DRRM processes resulted in the establishment of youth-managed DRRM resource centres in barangays Patao and Poblacion (Daanbantayan). These resource centres are equipped with educational materials and equipment to aid in risk communication, emergency kits and a physical space for meetings among community youth, both in- and out-of-school.

The children's dialogues may be considered a success in advancing children's right to participation as often children are deprived of spaces that allow them to articulate their views and even exact accountability for decisions that affect them.

The following were the lessons learned from the activities:

- In contrast to conventional approaches in governance, children's dialogues highlight the need for inclusiveness in agenda-setting and decision making. Children represent a largely marginalised sector in governance whose voices are rarely heard and whose actions are seldom recognised. The S3CDRRM Project demonstrated that children are capable of addressing 'serious' (read: adult) concerns and even of contributing to new creative ways of thinking and doing if given the opportunities to do so.

- Capacity-building activities with children held through a youth camp served to build children's confidence and enhance their skills in leadership, speaking up and involving themselves in governance. The youth camp was also a venue to nurture creativity which allowed children to voice their ideas through various means such as song, dance and visual art. This prepared children to engage with community stakeholders more effectively.

Good practice 4 – participatory 3D mapping: recognising children's roles in analysing risks

Participatory Community 3-Dimensional Mapping (P3D Mapping) is a recent innovation in mapping out hazards and vulnerabilities in various local settings. Using cheap and locally accessible materials, P3D Mapping enables collaboration among different community stakeholders and effectively bridges scientific and local knowledge. P3D Mapping consists of building scaled relief maps which are overlapped with other types of information such as terrain, land cover, infrastructure and location of at-risk groups. The superimposition of information on hazards, vulnerabilities and capacities allows for the concept of risk to become more tangible to local stakeholders. The process of building and updating the 3D map also facilitates a space for dialogue among community members and fosters ownership of risk data (UNISDR, 2014).

In a similar fashion, the S3CDRRM Project, in partnership with the University of the Philippines Department of Geography, facilitated the development of P3D maps in partner communities. Over a number of days, community members represented by a variety of sectors came together to build their 3D maps and overlay various risk information. Among the most notable observations made which contributed to the successful creation of barangay 3D maps was the active participation of children who were members of the Earth Savers' Club. Community youth leaders and other members of their respective organisations were at the forefront of building the 3D maps, and collecting and tabulating risk and vulnerability data. In this sense, children and youth meaningfully participated in a crucial DRR process that aimed to increase their respective communities' knowledge regarding the risks that they face.

The following were the lessons learned from the activities:

- The participation and initiative shown by children and youth proved to be crucial to the development of the 3D maps and contributed to the overall quality of the outputs. Thus, this experience demonstrates how mobilising children and youth can greatly enhance the development of a risk information system such as the 3D maps.

- Prior to the building of the maps, S3CDRRM Project staffs, youth leaders, and barangay workers set out to conduct an advance data validation and 'pre-plotting' session in order to expedite the process of building the maps. This initiative significantly cut down the amount of time it normally took to create the 3D maps.

Good practice 5 – children on the lead: supporting children to conceptualise, plan and implement disaster risk reduction initiatives

Adhering to the belief that children can be transformative agents in promoting disaster risk reduction and community resilience, the S3CDRRM Project mentored and provided financial support to organised youth groups to conceptualise and implement their own disaster prevention and mitigation activities, awareness-raising campaigns and capacity-building.

The Earth Savers' Club in Bantayan implemented three activities participated in by over 1,000 children and youth, local government officials, non-government organisation representatives, school administrators and local residents. Among the activities that the youth club organised was training on Water Search and Rescue in coordination with the local DRRM office. The youth leaders were taught basic emergency first responses – first aid and secondary care, basic search and rescue, and search and retrieval procedures in the water.

Meanwhile, in Daanbantayan, nine prevention and mitigation activities and five awareness-raising activities were implemented by the youth in communities. These activities were participated in by over 900 children, youth and local residents. Among the activities organised was storytelling for children emphasising the need for disaster risk reduction. Organised youth groups also conducted information campaigns in barangay Poblacion's various *sitios* (sub-villages) to popularise DRR messages. Other youth-led activities included mobile film showing campaigns, coastal clean-ups and mangrove planting activities.

The following were the lessons learned from the activities:

- Most child-focused DRRM still privilege adult voices in carrying out risk reduction planning and implementation. In contrast, the approach undertaken by S3CDRRM leveraged children's abilities to plan and conduct their own activities.
- The impact of this is not only felt among children who experienced a sense of empowerment, but also among a varied range of community stakeholders. Thus, a vital lesson learned is that investing in children's capacities and supporting them to take on their own initiatives brings benefits not only to the children but also to their communities.

Key takeaway learnings from the S3CDRRM Project

The S3CDRRM Project has demonstrated how Child-Centred DRR promotes the rights of children in DRR processes; but more importantly it also shows how investing in children also brings benefits to the wider community. Given the rich experiences and lessons learned from undertaking this one-year project, the following four points outline the major issues practitioners need to consider in developing DRR programmes.

Trust that children can take the lead in disaster preparedness and risk reduction. The most important obstacle to overcome is to change the way societies view children – as vulnerable, as incapable of making decisions, as powerless. By recognising children's agency, DRR practice can become a more effective and inclusive undertaking.

Invest in developing children's capacities to understand and analyse risks, communicate risk information and mobilise communities towards resilience. Children and youth have demonstrated that they can significantly contribute to resilience-building once they have access to the tools and knowledge that allow them to explore and use their skills to this purpose.

Support child-led activities and nurture children's creativity to develop and carry out plans and initiatives that contribute to disaster risk reduction. Children must be given enough space to explore their creativity and must feel empowered to undertake disaster risk reduction measures that benefit them, their families and their communities.

And lastly, *provide* an enabling environment for children to participate in DRRM processes. This requires gaining support from community institutions, including schools and local governments, in order to create spaces for children's meaningful participation. This should include institutionalising children's dialogues and ensuring youth representation in DRRM and local development councils. Indeed, this chapter provides an example of how EfS is practised beyond a classroom context. In sum, children do not need to be 'vulnerable' to disasters. They can be effective agents in building disaster resilience in their communities, through good practices of child-focused disaster risk reduction.

References

Back, E., Cameron, C., & Tanner, T. (2009). *Children and Disaster Risk Reduction: Taking Stock and Moving Forward.* Brighton: Institute of Development Studies. Retrieved 17 April 2017 from http://toolkit.ineesite.org/resources/ineecms/uploads/1057/Children_and_Disaster_Risk_Reduction.pdf.

Campbell, J., & Yates, R. (2006). Lessons for life: Building a culture of safety and resilience to disasters through schools – A briefing paper. Action Aid International. Retrieved 17 April 2017 from www.unisdr.org/2007/campaign/iddr/docs/UK-actionaid-report.pdf.

Hewitt, K. (1983). *Interpretations of Calamity from the Viewpoint of Human Ecology. The Risks and Hazards Series 1.* London and Sydney: Allen and Unwin Inc.

Mitchell, T., Haynes, K., Hall, N., Choong, W., & Oven, K. (2008). The roles of children and youth in communicating disaster risk. *Children, Youth and Environments Children and Disasters Children, Youth and Environments, 18*(181), 254–279.

Peek, L. (2008). Children and disasters: Understanding vulnerability, developing capacities and promoting resilience – An introduction. *Children, Youth and Environments, 18*(1), 1–29.

Plan UK. (2010). Child-centred disaster risk reduction building resilience through participation lessons from Plan International. Retrieved 17 April 2017 from www.plan-uk.org.

Tanner, T., Garcia, M., Lazcano, J., Tribunalo, B., & Seballos, F. (2009). Children's participation in community-based disaster risk reduction and adaptation to climate change. *Participatory Learning and Action, 60*, 54–64.

United Nations. (1990). Convention on the rights of the child. Retrieved 17 April 2017 from www.ohchr.org/EN/ProfessionalInterest/Pages/CRC.aspx.

United Nations International Strategy for Disaster Reduction (UNISDR). (2014). Participatory 3-dimensional mapping to foster multi-stakeholder collaboration in disaster risk reduction. The application to policy and practice. Retrieved 17 April 2017 from www.preventionweb.net/files/workspace/7935_cadaggaillardp3dm.pdf.

Wisner, P., Cannon, T., Davis, I., & Blaikie, B. (2004). *At Risk: Natural Hazards, People's Vulnerability and Disasters* (2nd ed.). London: Routledge.

4 Incorporating sustainability for general education

The challenge for large class teaching

Fa Likitswat

Introduction

We are living in a fast-moving world, with interconnected changes in environments, societies and economies, and globalisation. Lifestyle changes, fast consumption trends, urban lifestyles seem to develop and keep moving towards such unsustainable ways. One of the challenges that we face in education is how to incorporate sustainable concepts into the education of the younger generation and prepare them to deal with this complexity and these challenges? Sustainability as a word has an unclear meaning. It is even more challenging. In the educational field, policies and strategies must change in order to support the sustainability concept, and practice becomes concrete and visible.

In 2015, Thammasat University had a plan to restructure and redesign all of the general education core subjects. The main purpose for this development was creating the new curriculum for preparing the young generation to be able to adapt and get themselves ready for the global and local challenges related to the rapid change of environment, both social and economic. The old courses which had been taught for at least 25 years were replaced by new subjects, as well as new approaches to teaching and learning. The new courses were designed in terms of the world and the local communities facing challenges. Indeed, sustainability is one of the themes in one core course. The old subject which was called Integrated Technology and Science was taught by a lecture-based learning method. The new subject was developed and shifted the focus to Life and Sustainability guided by an active-learning approach. The new subject emphasises the dynamic interconnectedness of nature, consumerism and lifestyle, built-environment and energy consumptions as well as trends of global environment and Thailand's environment situation. It also reviews environmental conflicts and transformation towards sustainable lifestyles. Life and sustainability is divided into three modules; (1) environmental science, (2) built-environment, and (3) economic related issues.

In the fast-moving world that we live in, interconnected changes in environments, societies and economies, and globalisation take place rapidly. Lifestyle changes towards fast consuming trends are moving developments in unsustainable ways. In this resource consuming mainstream development, the ubiquitous existence of man-made objects, from the gigantic to microscopic, reflects human lifestyle that consumes and destroys the ecological pattern involving other species. The urgency for educational practice nowadays is on how to teach the next generation to be fully responsive, politically, socially, ecologically and historically, towards a sustainable way of living. To effect a major shift in attitude, we need to create more open-minded generalists who can see the complexity of connections instead of training more narrowed-minded specialists (Scott, 2000). Transforming general education courses to incorporate concepts and ways to practise sustainability presented challenging circumstances in a few ways.

Addressing sustainable concepts to the younger generations to prepare them to deal with complexity and new facing challenges is the key to the course's goal, but the meaning of the term sustainability in itself seems quite vast. It stands humbly as the keyword. The concept of sustainability and its implementation are still in the midst of development, thus communicating theoretical ideas as a clear and touchable concept to the younger generation is challenging. Educational policies and strategies need to be adjusted to help support the sustainability concept and practice to become concrete and visible. Although the concept has been widely introduced in Thailand, from the government, university, communities to a personal level, without a clear and focused plan, sustainability could not truly exist on the ground.

To create sustainable living, built-environments need major adjustments. Clearly, this is not only the task of planners, architects or engineers; it should involve a broader range of disciplines. According to Jucker (2002), the strategy of creating sustainable society has three important parts: (1) to point out unsustainable practices in the present contents and to identify its cause, (2) to identify base rules of which sustainable environments should be built, and (3) to evaluate which part of education has the potential to take its role in the transformation. Economic growth and human welfare seem to be in reverse correlation with sustainable concepts. Economic theory that aims for maximum profit from globalisation – the old way of thinking, does not help in this matter. The new paradigm needs to be redefined through learning from local perspectives and local knowledges.

> The role of general education is to make sure that students can learn broader concepts than their fields or expertise. Don't let students, who will become future decision makers, get stuck neither in their cultural vacuum or science and technological vacuum.
>
> (Cole & Merrill, 1982)

According to Cortese (2003), higher education has a crucial role in creating a sustainable future – to emphasise the designing of a sustainable human future with a new paradigm which shifts towards a systematic perspective, incorporating collaboration and cooperation. This emphasis should be a fundamental principle in all education. Estimating that the built-environment will double in size in the coming decades, a large impact of the built-environment on the human and natural world could become overwhelming. To incorporate sustainable concepts should be an urgent priority and it would be better addressed in large class-environments with diverse student backgrounds, through working in the field.

This chapter illustrates and reviews the course framework: how to incorporate sustainable learning as well as active-learning for the large classes at Thammasat University, using the built-environment module as an example. The built-environment module bridges cultural study, science and technology study that seems to have been lost in their own academic vacuums. The method of the study is to observe, analyse and discuss the teaching frameworks and students' learning outcomes by using the data collected throughout the 2015 academic year.

Frameworks for the sustainable built-environment

We create our living conditions by shaping the built-environment, while the built-environment itself influences the shaping of our way of life at the same time. To understand the changes that occur in this reciprocal relationship and the dynamics in built-environment is crucial because what we build has a significant impact on the natural environment upon which we depend. Those changes we make are affecting us both directly and indirectly. Climate change is a significant example. To learn about the built-environment could begin from the personal scale and then expand to a wider perspective: from the district to the city, and then the wider landscape. Students should discuss and raise questions on our way of living and the sustainable built-environment.

The sustainable built-environment module course objectives are: (1) to understand natural dynamics and how they affect the built-environment, (2) to realise the natural issues and how they impact on human lives, and (3) to raise questions and discuss about our lifestyles and how to create a sustainable future in built-environments. To create a new framework for Life and Sustainability, specifically for the built-environment module, some questions and issues frequently come up in committee discussion. The questions include the following:

- How to develop the course content to teach students without the architectural background?
- How to mobilise big classes into an active-learning environment?
- How to organise the teaching team?

- How to incorporate field study into the curriculum?
- How to demonstrate and interpret the complexity of sustainability reality and address its critical matter?
- How to shape the students to have critical minds and get ready to be strong decision makers with visionary perspective?
- How to stimulate and elevate sustainability design concept to the students in this chaotic physical world?

These questions were discussed and further developed to be the course framework as outlined below. These following issues were the fundamental frameworks for built-environment module: scale related sustainable built-environment contents and questions, teaching and learning team works, contextualisation by learning from the fields and shaping critical minds and strong decision makers focusing on physical designs and redevelopment solutions.

Scale related sustainable built-environment contents and questions

Course content in this module is divided into three topics related to the scale of the built-environment: human scale, landscape scale and city scale. We identified what the unsustainable factors of built-environment in each scale were by asking numerous questions like whether we live in sustainable buildings, neighbourhoods, communities, cities or landscapes. By breaking down the sustainable built-environment topics in relation to scales, students can see the direct and indirect relationships and impacts between themselves and unsustainable/sustainable physical environment distinctly.

For the human scale, the content involves the sustainable and unsustainable architectures, most of the time by questioning the appropriateness and compatibility of vernacular or international design of architectures locally and globally. Also, it provided the principle knowledge of the active versus passive design for the building energy consumption approach which is important for the real-life challenges.

At the landscape related scale, the concept of landscape as green infrastructure was adopted and explored. Since most of the cities were built on a flood plain, most cities are exposed to vulnerability to flooding conditions without realising that the wet conditions are quite normal and natural. Landscapes as green infrastructure increase the adaptability of the cities to those risks and play crucial roles in making the cities become more resilient. Landscape ecology is a hidden principle for this content – how to maintain the natural factors to co-exist in the built-environment patches. The paradigm of anthropocentrism versus ecocentrism is the main concept to be challenged showing a different way of thinking and living sustainably.

Finally, in the city related scale, the discourse illustrated the ability to cope with risks and natural disasters (e.g. earthquake). These structures reviewed the conflicts of our unsustainable lifestyles and the choices we have made to expand the built-environments, as well as the possibility to make the cities friendlier, supporting equality to the other living organisms and having the ability to cope with uncertain crises or disasters.

The class dynamics encouraged students to develop understandings of the concepts, models and prototypes of sustainable built-environments in each scale, using pop-up quizzes and short assignments. For example, in the human related scale, we discussed with the students the conditions of their homes as well as the opportunities to make their living using sustainable choices. For the landscape scale, questions about the function of green infrastructure such as rice fields or canals were asked and drew the connections of those multifunctional benefits to the human, social, economic, as well as for the cities. For the city related scale, the discussion on the vision of the future cities was the key focus. The class generated creative dialogues about sustainable built-environments with the hope that the students carry on these conversations with their families and outside communities.

Active-learning for classes of a big size

The course was attended by the freshmen of over ten diverse faculties, with 1,348 students (first semester) and 1,696 students (second semester). The process of active-learning allows not only the students but also the teaching team to develop their understandings of the dynamics and broad meanings of sustainability in each spatial built-environment and community. Applying new approaches of teaching and coaching with the emphasis on motivating the students to learn about sustainability, instead of providing information through lecture-based techniques. The challenge is how to make such a large class size with 300 to 700 students stay active. Thus, in each class, we divided the class into two parts: the first focuses on providing the content and principle baseline information on sustainability concepts in built-environments for the whole class, and the second part incorporated activities and workshops in small groups of about ten people. This way, students could focus on the content for short amounts of time, and then explore the concepts or ideas that they have learned in different assignments: mind mappings, quick sketches or essays. Flipping classrooms was another technique; we encouraged students to present and communicate about their works and share their ideas with the whole class.

Teaching and learning team work

Since the structure of the course is quite complex, we promoted the interdisciplinary teaching and learning experience. Collaboration between diverse disciplines plays a key part in completing sustainability projects. Students come from different backgrounds: social science, health science and technology sciences, including Faculty of Law, Business School, Faculty of Political Science, Faculty of Liberal Arts, Faculty of Economics, Faculty of Fine and Applied Arts, Faculty of Journalism and Mass Communication, Faculty of Public Health and Faculty of Science and Technology. There were three sections in each semester, run by the teaching team including faculty members and teaching assistants, mostly graduated architecture students. Following the university plan with one teaching assistant per 100 students (10 groups), we encourage the students to work across their disciplines.

Contextualisation, learning from the fields

Sustainability studies should not be only in the classrooms. Ideally, it should also be contextualised. The question is how to make large classes work with a place-based model. Since there was high student enrolment in both semesters, taking the entire class on an excursion would not be manageable. It brought on other questions in terms of transportation, communication and negative impact on the communities. Therefore, incorporating place-based learning into the group project assignments was the sustainable solution for going and working with communities in particular areas.

The purpose of the assigned projects was to incorporate local knowledge and external aspects, promoting all lifelong learning not only from the classrooms but from anywhere. In groups of ten students each students worked on their particular interests or topics related to spatial areas and specific issues. To identify interconnecting issues and highlight the critical points from the spatial areas, two distinct sets of questions were asked for both semesters. The first related to learning from landscape change and identifying important issues. The latter focused on living with disaster, adaptability and making resilient cities.

For the first semester, we created a list of places for students to use as an example of study areas. These could be local places, with different characteristics ranging from a historical town, commercial area, new development area and sub-urbanising area. While in the second semester, the study areas were widely open to the students to choose, since it would rely on natural disaster-related topics that could be both local and global in scale. There were weekly guidelines for students to submit their project work progress. They were developing content, creating mind-maps on cause and effect, framing the story telling techniques for their presentations, writing scripts or sets of questions, interviewing the communities or conducting focus group discussions in the study areas, then editing their short videos and compressing them to within the three-minute length.

Shaping critical minds and strong decision makers

To shape the analytical view points of the students requires the envisioning of critical minds. In the process of learning by doing, the projects shape the idea of teamwork, team building and generate dialogues between teaching teams, students and outside communities. Developing a special skill set to interact with society properly and creatively was important.

It is quite challenging to ask students without design backgrounds to design sustainable architectures, landscapes or cities. We literally ask the students to come up with the design concept or conceptualise the physical built-environment in their projects. In doing this we extend the common languages of 'design-built aspiration' for a better future – not only the aesthetic side, but on the sustainable and ecologically friendly side as well.

Learning outcome, finding and feedbacks

The learning outcomes from the students' projects in both semesters reflect different angles to our design and built-environment. With a limited time-frame, students had four to five weeks to work on their project and present their ideas in a three-minute video and written report.

Most of the projects presented environmental issues and conflicts as well as some of the social problems. Some of the projects reflected on how the students changed their attitudes about sustainability when they were working in the field. During the first semester, one of the groups chose Thaprachan, the area within the Rattanakosin historical district along the Choapraya River. In the beginning, students planned to work on the solid waste issues, but when they investigated the area and started to have conversations with local people, they realised that the footpath and walkway cleaning up programme were critical to the quality of life of people in that area. The students' final presentation captured the emotion and intention of the food sellers who had been doing this business for over 30 years. The questions related to the use of public landscape, public right of way, sanitation, urban gentrification and street food culture in Bangkok. For the second semester, one of the groups chose to work on a village in Ayutthaya, dealing with the flooding problem that occurs regularly throughout each year. The students discovered that the place they first planned to go was not the worst case among villages dealing with flooding issues, and the local people pointed towards a more critical area in Ayutthaya and the students subsequently based their project there. The interviews with the community reviewed the way to live with floods and reflected concepts of social responsibility and the relationship between rural areas and the city centre.

The study found the students' work presented some level of (1) collaboration between students in the team and the community, (2) ability to re-learn and re-value the critical and debatable issues in the communities, and (3) the process of learning and making decisions showing students' deep and critical analytical skills. The most surprising finding was that students had to learn from the process of working back and forth between their way of thinking and the observations as well as conversations with the communities. Their work requires further analysis in order to improve future assignment criteria that could help future students to learn better.

Education for sustainability (EfS) challenges

From direct feedback and our observation, the students reflected both positively and negatively. Some of the groups indicated that they enjoyed active-learning about new perspectives of sustainable built-environments, while other groups complained about the unclear purpose of general education, overlapping contents between each general education course, high workloads and their desire to study only in their fields and disciplines. These challenges are discussed below.

Why do I have to take this course?

Creating a meaningful learning experience is crucial. Since many students do not understand why they need to take this course, they feel that general education is unnecessary for their future career. They are eager to study in their majors or participate in their disciplines rather than working and studying with unfamiliar faculties or people from the different fields. Kirk-Kuwaye & Sano-Franchini (2015), in the article 'Why Do I Have to Take This Course?': How Academic Advisers Can Help Students Find Personal Meaning and Purpose in General Education', suggest that the academic advisor can also point out the deep meaning of general education values to undergraduate students. In the discussion for general education, we also had a discussion about adding a meaningful introduction to the university orientation camp.

New subject, new way of learning, lots of work to do

Since all of the new versions of general education courses were launched in 2015, there is some overlapping content and workload between each general education course. Some of the students made comments on the structure of similar content and assignment. Also mentioned was that there was too much work to do for those general education courses. Better organisation and high efficiency of education perspectives; we need better coordination between all levels of a teaching team in a particular course.

Eliminating the boundary between disciplines and broadening the student knowledge

Eliminating boundaries between disciplines challenges the class dynamics. Most students prefer to work with their friends or classmates who they know. It takes quite some encouragement for them to come out of their comfort zone and work with the other students from the different majors or disciplines. Building up their confidence to communicate and collaborate within diverse groups is crucial for creating sustainable communities. Common languages and some of the technical terms on built-environments were highlighted throughout the course structures, lectures and workshops to provide a broader perspective on built-environment and society.

Future planning and development

Challenges and difficulties were presented during the first two years of the course. The teaching team concluded that future planning and development for the next course iteration are as follows.

Promoting sustainability as valuable and meaningful learning

Emphasising the purpose of general education is needed in all levels of communication to shape the students' perspective and better understanding of the sustainability approaches. Effective communications are the key, working with all levels of collaboration, from personal, academic advisors, faculties, and university levels, promoting the meaningful purpose of general education as well as sustainability learning.

I – model

Instead of the traditional T model or up-side-down T model where the curriculum plan is for students to take general education courses either in their first or final year, I-model should be implemented for the learner to design their own schedule and register upon their will.

Partnership with the communities

Some areas have more potential to establish sustainable projects than other areas. These areas may already present issues related to sustainability and/or willingness to collaborate in working with the university. Pre-establishing these connections could help in accelerating the students' learning processes more efficiently within a short amount of time.

The next step

To teach and learn about sustainability concepts in such an unsustainable world is challenging. It requires a whole new way of thinking and the ability to re-learn from our failures. Large class sizes provided the opportunity for the teachers to appreciate diverse thinking, skills and backgrounds, although it also requires larger teaching teams to facilitate all students well. The well-balanced number of students should not exceed 200 per class. The findings from previous classes suggest further incorporation with the external communities. We need to create a sharp vision for a better future, not only for humans, but also for other organisms in the intricate web of interdependency.

In the introductory chapter, the editors suggest that it is important to develop agency among educators. Indeed, every teacher plays an important role in educating and bringing about changes in students' learning behaviour in both formal and informal educational settings. The discussion on large class sizes, community-based projects and innovative curriculum design are just some ways to enhance the role of the teacher in this endeavour. It is only through supporting teachers with these changes in the module design, and for a large class in this case, that we can hope to address issues about environments, societies and economies, and globalisation in a fast-changing world.

References

Boyatzis, R., Smith, M., & Blaize, N. (2006). Developing sustainable leaders through coaching and compassion. *Academy of Management Learning & Education, 5*(1), 8–24. Retrieved from www.jstor.org/stable/40212530.

Cole, H., & Merrill, W. (1982). Science, technology, and general education. *The Journal of General Education, 34*(3), 247–257. Retrieved from www.jstor.org/stable/27796915.

Cortese, A. (2003). The critical role of higher education in creating a sustainable future. *Planning for Higher Education, 31*, 15–22.

Jucker, R. (2002). 'Sustainability? Never heard of it!' *International Journal of Sustainability in Higher Education, 3*(1), 8–18. doi:10.1108/14676370210414146.

Kirk-Kuwaye, M., & Sano-Franchini, D. (2015). 'Why do I have to take this course?': How academic advisers can help students find personal meaning and purpose in general education. *The Journal of General Education, 64*(2), 99–105. doi:10.1353/jge.2015.0008.

5 Can grassroots organisations (GROs) replace government policy towards creating a sustainable climate change education programme in Singapore?

Andy Wi

Introduction

To create awareness about climate change, most governments employ a top-down approach, which is to provide as much information about climate change as possible. They hope that in doing so, people might take action for climate change and this approach has been successful in creating awareness among people in many countries (Abrahamse, Steg, Vleck, & Rothengatter, 2007; Chang, 2008). Thus, there is no shortage of pamphlets, brochures and letters provided by environmental agencies to remind and educate people about climate change management. However, studies in the literature have shown that increased awareness does not lead to action (Collins, Thomas, Willis, & Wilsdon, 2003), and having awareness about climate change is, in itself, not enough (Chang, 2014). Even when people have increased awareness, they might be uncertain about the new knowledge and not commit to any action due to the inconvenience to do so (Aarts, Paulussen, & Schaalma, 1997). Perhaps awareness is not a necessary factor but a pre-condition to climate change action.

Similarly, the Singapore government has also been promoting pro-environmental behaviour actively through campaigns and education (Chang, 2008; MEWR, 2008; Sustainable Singapore, 2015). It was through these measures that there was an increased awareness of the issue of climate change (NCCS, 2011). However, the question of why there is very little action taken despite the proliferation of information remains.

One of the problems is that people perceive it as a lack of empowerment to undertake actions that make a difference to climate change. Most of them think of approaches at the country level like implementing mitigation policies that encourage or mandate the use of solar panels, rooftop gardens and plastic recycling. All these require massive resources and skills to accomplish, which has consequently led to feelings of helplessness at the individual level. This feeling of helplessness together with the inability to act and mitigate climate change pushes them to see the government as the sole solution provider (Norgaard, 2009).

In addition, the massive media coverage over the last decade on climate change has left individuals wondering whether it can be done and whether it is the government's responsibility to manage climate change. Although there is a considerable effort by the mass media to educate the public about climate change, it has indirectly created the general perception that there is little an individual can do to stop climate change and it is perceived to be the responsibility of policymakers and non-governmental organisations (Chang, 2008; Norgaard, 2009). Given the people's dependence on the media to understand the problem, this barrage of information from multiple sources certainly does not help to promote climate change action. Coupled with the apparent denial and business-as-usual behaviour, it leaves individuals feeling neither guilt nor obligation to act (Carvalho, 2007; Chang, 2014).

Climate change is a global problem that requires social, technological and political relations to be successful (Räthzel & Uzzell, 2009), thus, the emphasis on either people's responsibility or the government's responsibility is not enough. It is not just for individuals to take action, nor is it the sole responsibility of policymakers. There is also a growing concern about the role of individuals in contributing to and bearing responsibility for climate change (Stern, 2000). The question of climate change then translates into who to attribute the problem to and whose onus it is to respond – the government's or the people's?

Or perhaps a third party – the grassroots organisation? In terms of creating climate change awareness and action, there are two key approaches: top-down and bottom-up (Wi, 2018).

Top-down approach

The top-down approach is generally used by most governments as an overall policy direction of the country (OECD, 2003). Climate change policies are always implemented and updated based on a top-down framework (OECD, 2003), as policymakers had always believed that they can limit the amount of emissions through a top-down governmental approach (N. Stern, 2009).

The reason for this is that governments are able to mobilise more resources and meet with the least resistance from the people (at least in Singapore). The main advantage of the top-down approach is the efficiency in making a decision (OECD, 2003), whereby directives are given at the higher levels to the lower levels in the hierarchy. This approach is most commonly deployed in the military or multinational companies.

In terms of environmental control and protection, the government has the full rein of control of the policy instrument and can choose between an incentive-based intervention or a control and command approach to intervention (Harrington & Morgenstern, 2007). Each of these has its pros and cons. In an incentive-based intervention, the government could encourage pro-environmental behaviour in its people by giving out incentives to those who adhere to the activity. This intervention is excellent for short term control but is not maintainable. Research has shown that as soon as the incentive is retracted, people would no longer follow through with the activity (Abrahamse, Steg, Vlek, & Rothengatter, 2005; Deci & Ryan, 2000; Geller, 2002; Stern, 2000). In the control and command intervention, the government could enforce taxes on over consumption, or laws, compliances and fines. One example of a successful implementation of this approach is the 'anti-littering' policy in Singapore, where a fine is imposed on anyone caught throwing rubbish in public areas.

However, one reason for the limited success in the top-down approach is that government officials are divided into different departments and divisions, having their own specialised task and specific mandates (Nijkamp & Perrels, 1994). Policymakers are also largely governed by competing priorities and interest groups, and make decisions with respect to their own agenda and schedules, and a clear inter-departmental communication is sorely absent (Wilson, 1989). They believe that there are many other policies more important than climate change itself, like inflation, unemployment, health issues such as dengue outbreaks, and food security (World Economic Forum, 2014). Moreover, in the majority of democratic countries where an election brings the government to power for only a period of four to five years, policymakers do not have the time to see a plan through. It takes a new policy a few years before the management plan comes into effect and the public will only be able to see any significantly visible results years after the implementation. At the point where there are some positive significant changes, it will most likely be towards the end of their term of office. As a result, projects are shelved and a new project takes its place. There is hence no long-term sustainability.

Bottom-up approach

This bottom-up approach is a consultative style of decision making that stems from the people (grassroots) upward (Wi, 2018). This approach emphasises community participation and grassroots organisations working together and leverages on knowledge and experiences of the people living in the community and creates a feeling of mutual ownership towards the activity/project. An example of the bottom-up approach is the 'community garden in Singapore' where the residents living in the area decide on how to grow, where to grow and what types of flowers to grow in the community garden (HDB, 2010). However, this approach lacks democratic control and there is difficulty in getting group consensus in making decisions (Fraser, Dougill, Mabee, Reed, & McAlpine, 2006).

Grassroots is defined in the Cambridge dictionary as ordinary people in a society or an organisation (Cambridge, 2018). They are mostly understood as a group of individuals who engage in community activities and interaction for the collective interest of improving their community (Wi, 2018). While grassroots organisations (GROs) and non-governmental organisations (NGOs) are similar in many ways (Dempsey, 2009), the former is the lowest level in which the NGOs operates (de Souza, 2007). NGOs may be national, multinational or transnational based on their operations and services (Lindenderg & Bryant, 2001), whereas GROs are more locally based (Dempsey, 2009; Eade, 2000). Turner and Hulme (1997) argue that 'complications' occur when some NGOs originate from local 'historical' settings and have been in existence for over a decade (p. 202). In general, NGOs are considered to be of a higher level and the differences between NGOs and GROs are in terms of the scale and the locality of the organisations.

The main objective of the GROs is to help people and to bring about changes in the community. In a grassroots community, communications and cooperation are important aspects for both individual and community level changes in building civil society and making a social change (Christens, 2010; Cole, 1974; de Souza, 2007). It means that individuals cooperate with one another in the community to achieve collective action and decision making. The grassroots community organising process strengthens commitment to civic involvement, and promotes volunteers' participation (Smith, 2000; Speer, Peterson, Zippay, & Christens, 2010). GROs also focus on local issues rather than national issues or advocating a particular group's interests (Mauzy & Milne, 2002). In Singapore, the RCs, the grassroots bodies present in Housing and Development Board (HDB) estates island-wide, are well placed to provide community support and help residents understand government policies (Mokhtar, 2014). GROs also set up neighbourhood meetings, which enhance citizen participation in the assessment of local needs and resources, and reinforce actions that build a sense of community (Pilisuk, MsAllister, & Rothman, 1996).

GROs have been successful in raising awareness and encouraging community involvement (Christens, 2010), due to the fact that they are closer to the community and understand the local context better (Ghai & Vivian, 2014; Mauzy & Milne, 2002). In addition, community-based approaches in local urban areas have significant benefits (Viswanathan et al., 2004) as they can serve as 'experimental laboratories' to test different solutions to public problems. These areas could also share information and findings while at the same time learning from others around the world (Kincaid, 1999). People working together with the GROs are not seen as individuals but as a collective unit that can bring local knowledge to inform policy makers (Christens, 2010; Wi, 2018). These small and discerning efforts in the immediate community can often have the most profound impact on people's lives. Successful communities' solidarity projects can often make more changes at a global level easier than individuals (Betsill, 2001). Of course, there remains the efficacy of local actions and the role of local governments in addressing the climate change phenomenon. While Wi and Chang (2018) have affirmed the role of GROs in influencing public policy, the success in Singapore could be due to its small area. However, this may be more difficult in larger countries where there are cultural, economic and social differences.

Which approach is more effective?

The top-down and bottom-up are just approaches to planning strategies. Top-down planning would typically cascade down from higher levels of governments while a bottom-up would start at the local level and progress upwards to the decision makers. Due to economies of scale for climate change mitigation, it is complementary and mutually beneficial for all levels of governments and community to be involved in it together (Deangelo & Harvey, 1998). Different countries might favour different approaches; in fact, most other countries have policies that have at least some element of both approaches (Harrington & Morgenstern, 2007).

Moreover, a government's policy or initiative would not be sustainable without the support of the people. Similarly, the people's initiatives would not have any impact without governmental support (Couto, 1998; Ghai & Vivian, 2014; Kelly & Caputo, 2006).

Wi (2018) argues that an integration of both approaches is necessary for climate change education and suggested the use of the Grassroots Approach (GRA) model. The GRA (see Figure 5.1) is a framework that describes the interrelationship between the government, grassroots and the people. It emphasises how the government, grassroots and people ideally work together to maintain a sustainable climate change programme/action.

It consists of six stages; namely policy planning, information and implementation, public consultations, involvement, feedback and advisory.

- **Policy planning** – to formulate policy at the governmental level
- **Implementation** – to implement and inform about the policy
- **Consultation** – consultation between the government and the people
- **Involvement** – the interaction and communication between the government, GROs and the people
- **Feedback** – a feedback channel for the people to communicate with the government
- **Advisory** – the government seeking advice from experts (e.g. private sectors or within government agencies)
- **Public Education Programme** (PEP) – providing training/education to the people (e.g. climate change education)

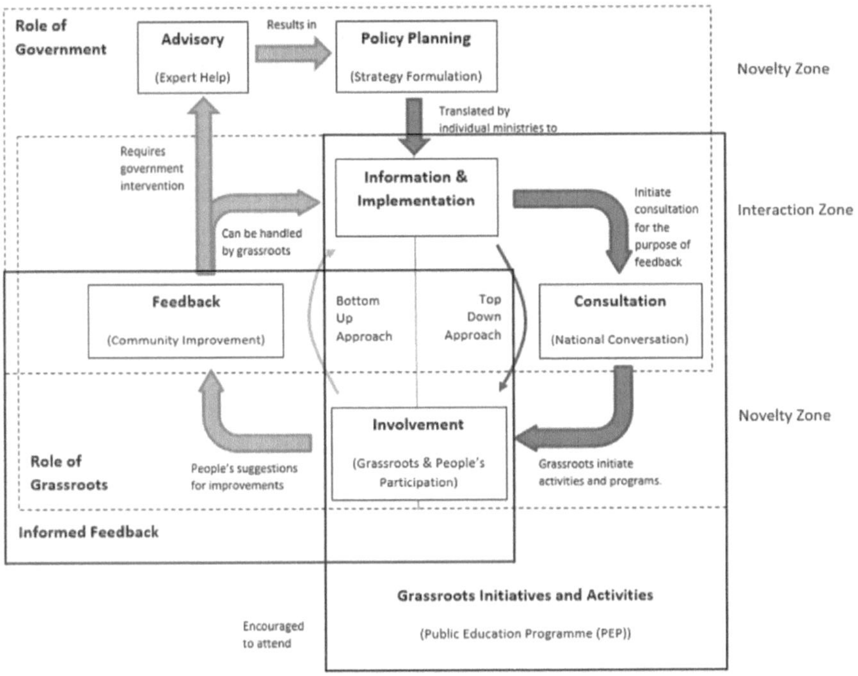

Figure 5.1 Grassroots Approach (GRA) model.
Source: Wi, 2018.

Based on the GRA approach, a study that was conducted in Singapore showed that residents who attended a specially developed climate change education programme and adopted the suggested pro-environmental behaviour were able to save approximately 25.6 kWh per month (Wi & Chang, 2018). Assuming that all 1,263,600 households in Singapore (Doss, 2016) adopt pro-environmental behaviour, the country could be saving 32.5 GWh of energy per month.

This approach is unique in that it shifts from the common established ideology of a top-down approach through a national level government, to society and then to the individual. It empowers individuals to adopt pro-environmental behaviour with the integration of both the top-down and bottom-up approaches through the support of GROs (Wi, 2018). For example, instead of just providing information about energy-saving tips, the GRO volunteers will explain the rationale of keeping the thermostat of the air conditioner at 25°C instead of just listing it as an energy-saving tip.

Where do we go from here?

Climate change is a global and collective problem and requires a collective solution in which responsible civic participation by individuals, communities and governments is necessary (Stoltman & Lidstone, 2001). Thus, the emphasis on either the government's or the people's responsibility is insufficient. At the national level, the government can come up with strategies and policies to manage climate change, but there is a limit to what the government can do about climate change. Apart from the fact that there are debates on whether human beings are at fault, the moral imperative requires people to do something about climate change. The main consumers of the resources depend on the people themselves and there is growing concern about the role of individuals in contributing to and taking responsibility for climate change (Stern, 2000). At the personal level, individuals must understand the climate change issue before they begin taking action for climate change (Chang, 2014). The ability to understand and take climate change action probably depends on where the person is from (e.g. a rural part of the region or a more developed area). While the environmental impact of any individual's personal behaviour is small, the impact on climate change becomes significant when many people independently do the same thing (Stern, 2000).

Climate change is happening whether or not we blame it on humans and is 'extremely likely' to be caused by people (IPCC, 2014). Our collective future is neither dependent on only a top-down governmental approach nor a bottom-up individual's approach, but on restoring and preserving the quality of the environment by living a pro-environmental lifestyle that could well be a product of both approaches. As Wi (2018) suggested, a good climate change education is not the sole responsibility of governmental institutions, the community and the people, but a collaborative effort among everyone (an integration between the top-down and bottom-up approach).

Bearing in mind that there is no 'one-size-fits-all' solution to combat climate change, it requires the collaborative efforts of the government, the community and the people (Wi, 2018; Adger, 2003; Ghai & Vivian, 2014). This is because climate change impacts different countries differently, in terms of environments and contexts. Thus, it will require different approaches to climate change education. At present, most climate change initiatives remain at the level of awareness building for many individuals. Therefore, to what extent people take action for what they understand and/or believe depends on where they are and how motivated they are.

Climate change is not an individual's problem or the sole problem of any country, but a common problem for Earth as well. Therefore, every country should play their part in the mitigation of climate change. As described in Hardin's (1968) Tragedy of the Commons, if individuals use common resources for their own gain and with no regard for others, all the Earth's resources will eventually be depleted. For a climate change education programme to be successful, it is important to understand an individual's perspective and the human-environment interaction.

Studies in the literature have shown that increased awareness does not lead to climate change action (Collins et al., 2003), and having awareness about climate change is in itself not enough (Chang, 2014). Even when people have increased awareness, they might experience uncertainty about the new knowledge and do not perform any action due to the feeling of inconvenience (Aarts et al., 1997). However, Poortinga, Steg, Vlek, and Wiersma (2003) suggest that pro-environmental behaviour is more acceptable if an individual understands that it benefits both the environment and the individual in terms of financial savings. Stern (2000) also argues that in recent years, many individuals are contributing to and bearing responsibility for climate change. Perhaps the responsibility technically lies in the hands of the people as organisations, companies and even governments are made up of people. This brings us back to the question of whose onus it is to educate the people about the climate change issue and empower them to take action – the government or the people?

GROs being closer to the community are better positioned to understand the local context, raise awareness and motivate the community to adopt pro-environmental behaviour (Florin & Wandersman, 1990; Ghai & Vivian, 2014; PA, 2011) through education and community support. In addition, GROs in different areas (e.g. estates, precincts) can modify and adapt different initiatives to the different demographics and cultures that suit the individuals living in the area. The GROs can educate the people regarding 'what to do?', 'how to do it?' and 'who will do what?' on climate change. Individuals working together with the GROs are seen as a collective entity that provide help, resources and opportunities for climate change action. Perhaps the responsibility of educating the individuals technically lies in the hands of the GROs; having the ability to integrate the government's top-down policies and people's bottom-up initiatives to climate change management.

Since it has taken governments around the world decades to come to a consensus on how best to tackle the climate change issue, perhaps it is time for the GROs to play a more evident role in educating and empowering the people to take action for climate change. At the local scale, the GROs can engage and empower individuals in different countries to take action in a way that is culturally and geographically efficient. At the global scale, these experiences can then be used to inform respective governments and lead the world in climate change action.

The Earth is our home and it is up to every individual, regardless of race, language or religion, to come together and take ownership of our shared planet Earth. The time to act for climate change is now. It is not just anyone's effort in particular but everyone's effort is needed collectively to engage our common environmental future. And as what the title states, let all take that step towards fighting climate change by being part of the grassroots organisations near their area. The philosopher Lao Tzu once said, 'Do the difficult things while they are easy and do the great things while they are small. The journey of a thousand miles begins with one step.'

References

Aarts, H., Paulussen, T., & Schaalma, H. (1997). Physical exercise habit: On the conceptualization and formation of habitual health behaviours. *Health Education Research, 12*(3), 363–374.

Abrahamse, W., Steg, L., Vlek, C., & Rothengatter, T. (2005). A review of intervention studies aimed at household energy conservation. *Journal of Environmental Psychology, 25*(3), 273–291.

Abrahamse, W., Steg, L., Vlek, C., & Rothengatter, T. (2007). The effect of tailored information, goal setting, and tailored feedback on household energy use, energy-related behaviors, and behavioral antecedents. *Journal of Environmental Psychology, 27*(4), 265–276.

Adger, W. N. (2003). Social capital, collective action and adaptation to climate change. *Economic Geography, 79*(4), 387–404.

Betsill, M. M. (2001). Acting locally, does it matter globally? The contribution of US cities to global climate change mitigation. Open meeting of the human dimensions of global environmental change research community, Rio De Janeiro, Brazil, 6–8.

Cambridge. (2018). Cambridge dictionary. Retrieved from https://dictionary.cambridge.org/dictionary/english/grassroots.

Carvalho, A. (2007). Communicating global responsibility? Discourses on climate change and citizenship. *International Journal of Media and Cultural Politics, 3*(2), 180–183.

Chang, C. H. (2008). *Climate and Climate Change: A Singapore Perspective.* Singapore: McGraw Hill.

Chang, C. H. (2014). *Climate Change Education: Knowing, Doing and Being.* Abingdon: Routledge.

Christens, B. D. (2010). Public relationship building in grassroots community organizing: Relational intervention for individual and systems change. *Journal of Community Psychology, 38*(7), 886–900.

Cole, R. L. (1974). *Citizen Participation and the Urban Policy Process.* London: Lexington Books.

Collins, J., Thomas, G., Willis, R., & Wilsdon, J. (2003). Carrot, sticks and sermons: Influencing public behaviour for environmental goals. A Demo/Green Alliance Report for Defra. Retrieved 5 January 2014 from www.demos.co.uk/files/CarrotsSticksSermons.pdf.

Couto, R. A. (1998). Community coalitions and grassroots policies of empowerment. *Administration & Society, 30*(5), 569–594.

Deangelo, B. J., & Harvey, L. D. (1998). The jurisdictional framework for municipal action to reduce greenhouse emissions: Case studies from Canada, USA and Germany. *Local Environment, 3*(2), 111–136.

Deci, E. L., & Ryan, R. M. (2000). The 'what' and 'why' of goal pursuits: Human needs and the self-determination of behaviour. *Psychology Inquiry, 11*(4), 227–268.

Dempsey, S. E. (2009). NGOs, communicative labor, and the work of grassroots representation. *Communication and Critical/Cultural Studies, 6*(4), 328–345.

Department of Statistics, Singapore (DoSS). (2014). Report on the household expenditure survey. Retrieved 1 January 2016 from www.singstat.gov.sg/docs/default-source/default-document-library/publications/publications_and_papers/household_income_and_expenditure/hes1213.pdf.

Eade, D. (Ed.). (2000). *Development, NGOs, and Civil Society*. Oxford: Oxfam.

Florin, P., & Wandersman, A. (1990). An introduction to citizen participation, voluntary organizations, and community development: Insights for empowerment through research. *American Journal of Community Psychology, 18*(1), 41–54.

Fraser, E. D., Dougill, A. J., Mabee, W. E., Reed, M., & McAlpine, P. (2006). Bottom up and top down: Analysis of participatory processes for sustainability indicator identification as a pathway to community empowerment and sustainable environmental management. *Journal of Environmental Management, 78*(2), 114–127.

Geller, E. S. (2002). The Challenge of Increasing Proenvironmental Behaviour. In Bechtel, R. B., & Churchman, A. (Eds.), *Handbook of Environmental Psychology*. New York: John Wiley & Sons, Inc.

Ghai, D., & Vivian, J. M. (2014). *Grassroots Environmental Action: People's Participation in Sustainable Development*: London: Routledge.

Hardin, G. (1968). The tragedy of the commons. *Science, 162*(1968), 1243–1248.

Harrington, W., & Morgenstern, R. D. (2007). *Economic Incentives Versus Command and Control: What's the Best Approach for Solving Environmental Problems? Acid in the Environment* (pp. 233–240). New York: Springer.

Housing Development Board, Singapore (HDB). (2010). Overview of community gardening. Retrieved 1 June, 2013 from www.hdb.gov.sg/fi10/fi10333p.nsf/w/CORHeartlandGreening?OpenDocument.

Intergovernmental Panel on Climate Change (IPCC). (2014). *Climate Change 2014: Mitigation of Climate Change. Contribution of Working Group III to the Fifth Assessment Report of the Intergovernmental Panel on Climate Change*. [O. Edenhofer, R. Pichs-Madruga, Y. Sokona, E. Farahani, S. Kadner, K. Seyboth, A. Adler, I. Baum, S. Brunner, P. Eickemeier, B. Kriemann, J. Savolainen, S. Schlömer, C. von Stechow, T. Zwickel, & J. C. Minx (Eds.)]. Cambridge and New York: Cambridge University Press.

Kincaid, J. (1999). The international competence of US states and their local government. *Regional and Federal Studies, 9* (Spring), 111–133.

Kelly, K., & Caputo, T. (2006). Case study of grassroots community development: Sustainable, flexible and cost-effective responses to local needs. *Community Development Journal, 41*(2), 234–245.

Lindenberg, M., & Bryant, C. (2001). *Going Global: Transforming Relief and Development NGOs, 220*. Bloomfield, CT: Kumarian Press.

Mauzy, D. K., & Milne, R. S. (2002). *Singapore Politics under the People's Action Party*. London and New York: Routledge.

Ministry of Environment and Water Resources, Singapore (MEWR). (2008). Singapore's national climate change strategy. Retrieved 18 January 2012, from http://app.mewr.gov.sg/data/ImgUpd/NCCS_Full_Version.pdf.

Mokhtar, M. (2014). RCs can help link up residents: PM. *The Sunday Times*. Retrieved 1 September 2015 from http://ifonlysingaporeans.blogspot.sg/2014/03/rcs-can-help-link-up-residents-pm.html.

National Climate Change Secretariat, Singapore (NCCS). (2011). Public perception survey 2011, annex b. National Climate Change Secretariat, Singapore. Retrieved 20 November 2012 from http://app.nccs.gov.sg/data/resources/docs/AnnexB.pdf.

Opdenakker, R. (2006). Advantages and disadvantages of four interview techniques in qualitative research. Paper presented at the Forum Qualitative Sozialforschung/Forum: Qualitative Social Research.

Organisation for Economic Co-operation and Development (OECD). (2003). *Governance of Public Research Toward Better Practices: Toward Better Practices.* London: OECD Publishing. Retrieved from www.oecd-ilibrary.org/science-and-technology/governance-of-public-research_9789264103764-en.

Peoples' Association, Singapore (PA). (2011). About grassroots organisations. Retrieved 2 October 2012 from www.pa.gov.sg/about-grassroots-organisations.html.

Pilisuk, M., McAllister, J., & Rothman, J. (1996). Coming together for action: The challenge of contemporary grassroots community organizing. *Journal of Social Issues, 52*(1), 15–37.

Poortinga, W., Steg, L., Vlek, C., & Wiersma, G. (2003). Household preferences for energy-saving measures: A conjoint analysis. *Journal of Economic Psychology, 24*(1), 49–64.

Räthzel, N., & Uzzell, D. (2009). Changing relations in global environmental change. *Global Environmental Change, 19*(3), 326–335.

Smith, D. H. (2000). *Grassroots Associations.* Newbury Park, CA: Sage Publications.

Souza, R. T. de (2007). *NGOs and Empowerment: Creating Communicative Spaces in the Realm of HIV/AIDS in India.* Indiana: Purdue University.

Speer, P., Peterson, N., Zippay, A., & Christens, B. (2010). Participation in congregation-based organizing: Mixed-method study of civic engagement. Using evidence to inform practice for community and organizational change (pp. 200–217). Retrieved from https://pdfs.semanticscholar.org/8958/4b60c3f90fb3b1e1c530c632a23a5c0f8a00.pdf.

Stern, N. (2009). A Global Deal on Climate Change. In Saw, S. & Quah, D. (Eds.), *The Politics of Knowledge* (pp. 111–155). London: Institute of Southeast Asian Studies.

Stern (2000). Toward a coherent theory of environmentally significant behavior. *Journal of Social Issues, 56*(3), 407–424.

Stoltman, J. P., & Lidstone, J. (2001). Citizenship education: A necessary perspective for geography and environmental education. *International Research in Geographical and Environmental Education, 10*(3), 215–217.

Sustainable Singapore. (2015). Sustainable singapore blueprint. Retrieved 1 October 2015 from www.mewr.gov.sg/ssb/files/ssb2015.pdf.

Turner, M., & Hulme, D. (1997). *Governance, Administration, and Development: Making the State Work.* West Hartford, CT: Kumarian Pr Inc.

Viswanathan, M., Ammerman, A., Eng, E., Garlehner, G., Lohr, K. N., Griffith, D., ... Lux, L. (2004). Community-based participatory research: Assessing the evidence. Rockville (MD): Agency for Healthcare Research and Quality (US); 2004 Jul. (Evidence Reports/Technology Assessments, No. 99.) Retrieved from www.ncbi.nlm.nih.gov/books/NBK37284/.

Wi, A. (2018). Citizen participation as a key enabler for successful public education policies in climate change mitigation in Singapore. *International Research in*

Geographical and Environmental Education. doi:10.1080/10382046.2018.1430789

Wi, A., & Chang, C. H. (2018). Promoting pro-environmental behaviour in a community in Singapore – From raising awareness to behavioural change. *Environmental Education Research*. doi:10.1080/13504622.2018.1528496.

Wilson, J. Q. (1989). *Bureaucracy: What Government Agencies Do and Why They Do It*. London: Basic Books.

World Economic Forum. (2014). Top 10 trends of 2014. Outlook on the global agenda 2014. Retrieved 8 August 2015 from http://reports.weforum.org/outlook-14/top-ten-trends-category-page/.

Part II

What does an EfS curriculum look like?

6 The question of 'knowledge' about disaster risk reduction in sustainability education

Liberty Pascua

Introduction

The United Nations Office for Disaster Risk Reduction (UNISDR, 2009) defines a disaster as:

> a serious disruption of the functioning of a community or a society involving widespread human, material, economic or environmental losses and impacts, which exceeds the ability of the affected community or society to cope using its own resources. This definition highlights the multiplicity of factors that result in disasters such as exposure to hazard, vulnerabilities and inadequate coping capacities. The consequences of disasters include one or more of the following: loss of life, injury, disease and other negative effects on human physical, mental and social well-being, together with damage to property, destruction of assets, loss of services, social and economic disruption and environmental degradation.
>
> (UNISDR, 2009, p. 9)

Hazards and related disasters are integral to all three pillars – environment, social, economy – of the sustainability paradigm. In *environmental* sustainability, the occurrence of hazards and disasters have been known to devastate, regenerate and transform the ecologies and natural habitats of living things. In *social* sustainability, research shows that exposure to disruptive events could upset and disband established social networks (Manfredi et al., 2014), as well as reconfigure how these social networks are formed (Tobin, Whiteford, Murphy, Jones, & McCarty, 2014). In *economic* sustainability, disasters brought about by hazards are recognised as the variable that disrupts affected economies, rendering already vulnerable populations to a state of severe fiscal insecurity. With the rapid growth of urbanisation, the liveability of cities are becoming front and centre in the discourse of sustainable development (Asprone, Prota, & Manfredi, 2014; Gasparini, Di Ruocco, & Russo, 2014; Manfredi et al., 2014).

It is then imperative for sustainability education to be attentive to hazards and disasters in its coverage. This is true especially in considering the types of knowledge that teachers have to teach and students have to learn. The implications on teacher education in EfS cannot be understated. UNESCO's Education for Sustainable Development (ESD) programme, for instance, emphasises the centrality of building awareness and knowledge about and capacities for individuals and communities in view of present and projected calamities that threaten the sustainability of development.

Policy basis for disaster risk reduction (DRR) education

In recognition of the immense impact that hazards and disasters have on people and economies of the world, the United Nations and its affiliated agencies have come up with a series of policy initiatives, enjoining governments and non-state actors to work together towards the mitigation and improved adaptation of communities to risks associated with hazards and disasters. The years 1990–1999 marked the International Decade for Natural Disaster Prevention (IDNDP), the aim of which was to address through concerted international efforts the loss of life, poverty exacerbation and disruptions to social and economic processes (United Nations, 1999) caused by disasters. The first World Conference on Disasters took place within this period in Yokohama, Japan. The Yokohama Strategy was drafted, stipulating the need for countries to make a political commitment to reducing the vulnerability of their citizens (United Nations, 1994).

Action in that decade was succeeded by the implementation of the International Strategy for Disaster Reduction 2000–2007 (United Nations, n.d.). Building upon the experience from IDNDP, the United Nations Office of International Risk Reduction then shifted its focus from disaster response to disaster reduction. In 2005, the Second World Conference on Disasters took place in Hyogo, Japan. This was a historical event such that the Hyogo Framework for Action (HFA), the first global platform on DRR, was adopted and promulgated. Covering yet another decade of implementation (2005–2015), the HFA focused on policies aimed at building the resilience of communities, paving the way for greater participation of national governments in the DRR discourse through ministerial conferences (United Nations, 2005).

The Sendai Framework for Disaster Reduction (SFDRR), unveiled at the Third World Conference on Disasters in Sendai, Japan, succeeded the HFA. Building on the momentum created by its predecessor in raising awareness and generating political commitments for DRR, the SFDRR sets the 15-year roadmap of renewed priorities, specifically targeting 'the substantial reduction of disaster risk and losses in lives, livelihoods and health and in the economic, physical, social, cultural and environmental assets of persons, businesses, communities and countries' (United Nations, 2015, p. 12).

Disaster management evolved from rhetoric of prevention and preparedness, disaster reduction and finally focusing on risk reduction related to disasters (Chatterjee, Shiwaku, Das Gupta, Nakano, & Shaw, 2015) in the three world conferences (Yokohama 1994; Kobe 2005; Sendai 2015) and the corresponding frameworks of cooperation and strategies crafted and inked in these venues. Knowledge of disasters is a recurring theme – a central element in the HFA and the SFDRR. In its five-point agenda, for instance, the HFA underscores 'knowledge, innovation and education' (United Nations, 2005, p. 9) as an indispensable element in attaining its goal of building resilient communities. It calls on governments to prioritise DRR integration in their formal, informal and non-formal education channels (United Nations, 2005). The SFDRR reiterated the same principle. First in its four-point priorities for action is its emphasis on building an understanding of disaster risk. Included among the SFDRR's strategies is the incorporation of disaster risk knowledge in formal and non-formal education at all levels (United Nations, 2015). Advocating for a multi-hazard perspective, the framework acknowledges the shortcomings of DRR as implemented in the past, highlighting the lack of knowledge integration at the community, regional and national levels (Weichselgartner & Pigeon, 2015).

Knowledge and DRR

Knowledge saves lives. This is the premise of both the HFA and SFDRR's call to governments to integrate the learning of DRR in every country's education system. On the type of knowledge that should be taught, the literature is conflicted in regard to knowledge and learning for DRR. At one end of the spectrum are the optimists, who view knowledge almost as a panacea. Occupying the middle ground are the cynics, who point to the promises and failures of knowledge, owing to its politicisation and negligence in acknowledging crucial elements in closing the gap between behaviour and desired action. On the extreme end are the realists/pessimists who, upon reviewing patterns of learning and adaptation of populations from centuries past, conclude that societies are unable to learn from disasters.

Knowledge is seen by optimists as a potent catalyst in the reduction of risk. Well-informed people, it is posited, develop a culture of prevention and resilience in the face of disasters (United Nations, 2005). The optimists place significant confidence in the capacity of knowledge to aid in the minimisation of fatalities and economic losses caused by disasters. It is believed that those with comprehensive understanding of risks will act rationally; they will avoid precarious environments, put into place structures and measures to secure properties and livelihoods, be able to prepare for the worst and resilient enough to bounce back in the aftermath (Marincioni, 2007; McGinn, 1985; Rattien, 1990; Reiff, 1985; Romo-Murphy, James, & Adams, 2011). It is maintained that individuals and organisations are empowered to proactively take charge of their well-being and that of their communities when they know exactly what disasters are and what they can do to avert risks and losses. They deem that populations exposed to natural disasters, those with developed disaster memories, develop coping mechanisms that enable them to successfully mitigate disasters and adapt. Hence, proponents of this belief also put considerable faith in learning as a vehicle to promote deeper understanding of disasters (Kelman, Petal, & Glantz, 2015; Petal, 2015).

The increasing statistics on disaster-related losses over the past centuries despite the abundance of, and improved access to, information, casts doubt on the claims of knowledge as a catalyst for risk reduction. In the words of White, Kates and Burton, the knowledge-results gap shows that collectively, the world is 'knowing better and losing even more' (2001, p. 81). Indeed, knowledge is an area of contention among DRR scholars. Questions such as what and whose knowledge is cultivated, ignored and suppressed give rise to the issue of representation of voices and rights in the DRR discourse. Distinctively, there is a critical collective overtly opposing technocratic approaches that favour the use of science-based knowledge in creating policies and programmes of development without due consideration of indigenous perspectives and local customs (Gaillard & Mercer, 2013; Iloka, 2016; Kelman, Mercer, & Gaillard, 2012; Mercer, Kelman, Suchet-Pearson, & Lloyd, 2009; Mercer, Kelman, Taranis, & Suchet-Pearson, 2010; Walshe & Nunn, 2012). It is posited that the politicisation of knowledge, of the conscious or inadvertent dismissal of the social and cultural properties of understanding, has led to unintended consequences (Spiekermann, Kienberger, Norton, Briones, & Weichselgartner, 2015), adding several layers of complexity to the whole gamut of issues surrounding disaster management. One solution put forward is a conscious and democratic coproduction of knowledge including all stakeholders (Spiekermann et al., 2015; Weichselgartner & Pigeon, 2015).

On the extreme end, realists dismiss declarations on the potency of knowledge for risk reduction. Studies that assessed societies' ability to learn, mitigate and adapt to disasters indicate that disaster memory could only last a generation or two. People forget, they move places, even consciously wiping away disaster memory along with the trauma associated with hazards (Aguirre & Best, 2015; Bründl, 2015; Lübken, 2015; Pfister, 2015). Scholars who share this perspective argue that the confidence afforded to knowledge and learning, with emphasis on resilience, should be tempered in favour of reliable systems (Aguirre & Best, 2015).

There is much more positive regard to knowledge in the realm of education and learning, although the contradictions linked to DRR knowledge construction are also very much manifest. The literature shows that there is a concerted effort to integrate DRR in formal education (Apronti, Osamu, Otsuki, & Kranjac-Berisavljevic, 2015; Baytiyeh, 2014; Baytiyeh & Naja, 2014; Galliara & Prabhawalkar, 2012; McGinn, 1985; Rivera & Miller, 2009; Shaw, Mallick, & Takeuchi, 2011; Shaw, Takeuchi, &

Shiwaku, 2011; Shiwaku & Fernandez, 2011; Soffer, Goldberg, Avisar-Shohat, Cohen, & Bar-Dayan, 2010).

Some studies lament the sparseness of coverage in the curriculum of specific subjects (Clary, 1996; Tait, 1996; Whitehead, 1996). Others point to the lack of issues-based discussions (Okpala, 1996) and inattention to the need to address the knowledge-behaviour gap (Kanchev & Tsankova, 1996; Lidstone, 1996). Resonant with the conflicts inherent in knowledge construction, scholars are pushing for indigenous knowledge(s) to be given room and emphasis in education delivery. A democratic, rights-based inclusive approach is advocated, taking into account the perspectives of stakeholders forming a bottom-up and top-down approach to DRR education (Hiwasaki, Luna, Syamsidik, & Shaw, 2014; Mudavanhu et al., 2015; O'Hern & Nozaki, 2014).

The place of place-based knowledge

In this chapter, I take into account the different perspectives afforded to knowledge in arguing for a place-based orientation for DRR education. I refer mostly to the work of DRR researchers using the cultural approach to risk and disaster as my springboard in arguing for a postcolonial calibration of mainstream approaches to DRR education, specifically in areas where the hazards and related disasters are seen as 'chronic' (Tobin et al., 2014), 'normalized' (Bankoff, 2004; van Voorst, Wisner, Hellman, & Nooteboom, 2015), 'frequent life experience' (Bankoff, 2007, p. 26). I digress from a 'disaster lens perspective' (van Voorst, 2016), a [Western]-oriented epistemological analysis (Bankoff, 2010; Heijmans, 2009; van Voorst, 2016) whose view of risk events typically revolves around the idea of being abnormal and irruptive without due regard to larger context that include complex layers of risk and normalised hazards. Rather, I put forward the idea that hazards and related disasters are risk events that are not necessarily non-routine and are, in fact, enmeshed in people's perennial realities as predictable occurrences.

Postcolonial theories' appeal lies in their interrogation of ethnocentrism (Alatas, 1972; Andreotti, 2011), posing a sustained challenge on claims to the universality of dominant epistemologies (Tikly & Bond, 2013), and insistence on the plurality of knowledge (Kayira, 2015). As a school of thought, postcolonial theories emerged to resist the empire (Go, 2016); they locate European colonisation and expansionism as the main referent for discursive analysis, linking present-day power struggles, enduring inequalities, repressive practices and dislocated identities to the colonial experience. Politically, the collective tenets of postcolonial thought bring to the fore the limits of Western/Enlightenment thinking (Andreotti, 2011). Through a deconstruction of Western-centric perspectives, they show the sustained and ambivalent effects of colonial power structures on knowledge production (Go, 2016) engaging its scholars in active opposition to the varied manifestations of epistemological violence by pushing for non-Western modes of discourse (Alatas, 1972; Quayson, 2000).

Bankoff's thick description of a 'cultures of disaster' provides accounts of how, for instance, local architecture (Bankoff, 2015a), agricultural practices (Bankoff, 2003) and belief systems (Bankoff, 2015b) of communities are knitted together in the larger patchwork of constant exposure to hazards. Whereas inhabitants in these conditions are often portrayed as victims of the elements, studies show that the opposite is true; people and communities living with the constant threat of disasters are active survivors (Rodríguez, Quarantelli, & Dynes, 2007) whose knowledge of hazards and disasters, both practical and philosophical, are invaluable resources for DRR. They have developed and advanced ways and measures to cope with and survive, building cultures of disaster (Bankoff, 2003), with their concomitant base knowledge, which is practical and philosophical, local and universal, passed on from one generation to the next. Among the recommendations by Manfredi et al. (2014) in fostering resilience to ensure sustainability in the face of hazards and disasters are for successful local resilience experiences to be transformed into long-run adaptive practices. In addition, Manfredi and company also recommended that disaster risk knowledge should be increased among individuals and communities to contribute to resilience.

Conclusion

There is faith in knowledge and what it can do for DRR. The literature is indicative of different viewpoints on the uses of knowledge for the reduction of risks in sustainability education. However, the official knowledge developed in schools is shaped mainly by the dominant techno-scientific epistemology effectively stunting the co-construction of plural knowledge and rendering the field of DRR education denied the contribution of equally relevant epistemologies. There exists a wealth of knowledge that is specific to precarious environments and the people inhabiting it. DRR education should tap into this.

Research on the topic is defined by categories such as indigenous, scientific, school-based and community-based among others. It is posited that such neat categorisation fails to acknowledge the interaction between knowledge and the changes that happen to each knowledge stock over time. There are instances of crossovers and overlaps, but the direction of inquiry commonly begins by showing partiality for one type of knowledge as dependent on the academic orientation of the authors. Moreover, the assignment of knowledge into silos renders a privileged few to be given emphasis while others become subdued, neglected or ignored. To insist on privileging one type of epistemology, to be dismissive of explanations to a phenomenon that diverge from mainstream knowledge, inevitably discards valuable information critical in the occasion of hazard-induced disaster.

These artificial compartments to knowledge only perpetuate the universalist claims of dominant knowledge and the stereotypes afforded to non-mainstream epistemologies. Research in this area should draw attention to dominant representations and linear views that marginalise and discount places and people in manners that may be intentional or inadvertent. This requires a rethinking of methods and analytical lenses, of considering not just 'Western' and 'Southern' ways of seeing, but more so for allowing local, context-specific epistemologies to emerge. Operationally, this requires an in-depth understanding of the context of learning in disaster prone areas. It also means a holistic examination of top-level policies vis-à-vis ground-level interpretation and adaptation of said policies, symptoms of resistance and appropriation to technocratic rule, as well as the ambivalences of people in navigating the linkages of tradition and modernity in understanding their everyday experiences in disaster contexts.

This chapter presents a core conviction for disaster risk education as part of sustainability education: to reject artificial knowledge compartments and to adopt a view of hazards and disasters as fundamentally context-bound. A place's geography, politics, economics and social organisations are among key elements that define the severity of hazards' impacts and the vulnerability of affected people. As a single hazardous event affects one locality differently from another, it follows that the knowledge stock that is efficacious in one place may be pointless in the other. It then asserts for an approach that should not only be cognizant of place-based epistemologies but should be integral to the content and form of education. Educators should be wary of falling behind the agenda of dominant academic blocs and be critical of mainstream epistemologies that create artificial compartments of knowledge, expediently ignoring the applicability of (re)presented information to a place, its geography and its people. The idea that place-based epistemologies are static, backward, unscientific and irrelevant should be rejected. Instead, place-based epistemologies evolve. It is science with respect to the local climate and local traditions. It is geography that analyses the local terrain. It is history with an understanding of how hazards have shaped the story of a place.

DRR as a subset of sustainability education should transcend the ethno- and discipline-centric silos in research and practice, refuse the essentialist views about place-based epistemologies, and work towards fostering an educational encounter that empowers the learner, the teacher and their communities in adapting to life punctuated by hazards and disasters. Research and practice in DRR for sustainability education should then examine the representation of knowledge in mainstream DRR education. Such engagements should be most attentive to the voices and agencies of actors as (un/under/mis/over)represented in DRR learning in different education channels, actively deconstructing the politics inherent in DRR knowledge construction, with the purpose of generating informed recommendations to better harness policies and practices in DRR learning.

References

Aguirre, B. A., & Best, E. (2015). How Not to Learn: Resilience in the Study of Disaster. In Egner, H., Schorch, M., & Voss M. (Eds.), *Learning and Calamities: Practices, Interpretations, Patterns* (pp. 216–232). New York: Routledge.

Alatas, S. H. (1972). *Modernization and Social Change: Studies in Modernization, Religion, Social Change and Development in South-East Asia*. Sydney: Angus & Robertson.

Andreotti, V. (2011). *Actionable Postcolonial Theory in Education*. New York: Palgrave Macmillan.

Apronti, P. T., Osamu, S., Otsuki, K., & Kranjac-Berisavljevic, G. (2015). Education for disaster risk reduction (DRR): Linking theory with practice in Ghana's basic schools. *Sustainability, 7*(7), 9160–9186. doi:10.3390/su7079160.

Asprone, D., Prota, A., & Manfredi, G. (2014). *Linking Sustainability and Resilience of Future Cities. In Resilience and Sustainability in Relation to Natural Disasters: A Challenge for Future Cities* (pp. 55–65). Cham: Springer.

Bankoff, G. (2003). *Cultures of Disaster: Society and Natural Hazards in the Philippines*. New York and London: Routledge.

Bankoff, G. (2004). In the eye of the storm: The social construction of the forces of nature and the climatic and seismic construction of God in the Philippines. *Journal of Southeast Asian Studies, 35*(1), 91–111.

Bankoff, G. (2007). Comparing vulnerabilities: Toward charting an historical trajectory of disasters. *Historical Social Research/Historische Sozialforschung*, 103–114.

Bankoff, G. (2010). No such thing as natural disasters. *Harvard International Review, 23*, n.p.

Bankoff, G. (2015a). Design by Disasters: Seismic Architecture and Cultural Adaptation to Earthquakes. In Kruger, F., Bankoff, G., Cannon, T., Orlowski, B., & Schipper, E. L. F. (Eds.), *Cultures and Disasters: Understanding Cultural Framings in Disaster Risk Reduction* (pp. 53–71). Abingdon, Oxon: Routledge.

Bankoff, G. (2015b). Learning about Disasters from Animals. In Egner, H., Schorch, M., & Voss, M. (Eds.), *Learning and Calamities: Practices, Interpretations, Patterns* (pp. 42–55). New York: Routledge.

Baytiyeh, H. (2014). How can school education impact earthquake risk reduction in Lebanon? *Education, Business and Society: Contemporary Middle Eastern Issues, 7*(2/3), 120.

Baytiyeh, H., & K. Naja, M. (2014). Can education reduce Middle Eastern fatalistic attitude regarding earthquake disasters? *Disaster Prevention and Management, 23*(4), 343–355. doi:10.1108/DPM-12-2013-0219.

Bründl, M. (2015). Analyses of Natural Disasters and Their Contribution to Changes in Natural Hazard Management in Switzerland. In Egner, H., Schorch, M., & Voss, M. (Eds.), *Learning and Calamities: Practices, Interpretations, Patterns* (pp. 199–215). New York: Routledge.

Chatterjee, R., Shiwaku, K., Das Gupta, R., Nakano, G., & Shaw, R. (2015). Bangkok to Sendai and beyond: Implications for disaster risk reduction in Asia. *International Journal of Disaster Risk Science, 6*(2), 177–188. doi:10.1007/s13753-015-0055-4

Clary, M. (1996). Teaching about natural hazards and disasters in French secondary schools. In Lidstone, J. (Ed.), *International Perspectives on Teaching about Hazards and Disasters* (pp. 39–46). Clevedon, UK: Channel View Publications.

Galliara, M., & Prabhawalkar, A. (2012). Disaster management and role of academic institutions. *Social Work Chronicle, 1*(1), 1.

Gaillard, J. C., & Mercer, J. (2013). From knowledge to action: Bridging gaps in disaster risk reduction. *Progress in Human Geography, 37*(1), 93–114. doi:10.1177/0309132512446717.

Gasparini, P., Di Ruocco, A., & Russo, R. (2014). Natural Hazards Impacting on Future Cities. In *Resilience and Sustainability in Relation to Natural Disasters: A Challenge for Future Cities* (pp. 67–76). Cham: Springer.

Go, J. (2016). *Postcolonial Thought and Social Theory*. New York: Oxford University Press.

Heijmans, A. (2009). The social life of community-based disaster risk reduction: Origins, politics and framing. Disaster Studies Working Paper No. 20. Aon Benfield UCL Hazard Research Center, London, UK.

Hiwasaki, L., Luna, E., Syamsidik, & Shaw, R. (2014). Process for integrating local and indigenous knowledge with science for hydro-meteorological disaster risk reduction and climate change adaptation in coastal and small island communities. *International Journal of Disaster Risk Reduction, 10*, 15–27. doi:10.1016/j.ijdrr.2014.07.007.

Iloka, N. G. (2016). Indigenous knowledge for disaster risk reduction: An African perspective. *Jàmbá Journal of Disaster Risk Studies, 8*(1), 1–e7. doi:10.4102/jamba.v8i1.272.

Kanchev, D., & Tsankova, L. (1996). The Elemental Natural Phenomena in the Geography Teaching in Bulgarian Schools. In Lidstone, J. (Ed.), *International Perspectives on Teaching about Hazards and Disasters* (pp. 47–58). Clevedon, UK: Channel View Publications.

Kayira, J. (2015). (Re)creating spaces for uMunthu: Postcolonial theory and environmental education in southern Africa. *Environmental Education Research, 21*(1), 106–128.

Kelman, I., Mercer, J., & Gaillard, J. C. (2012). Indigenous knowledge and disaster risk reduction. *Geography, 97*(1), 12–21.

Kelman, I., Petal, M., & Glantz, M. (2015). Using Spare-time University for Disaster Risk Reduction Education. In Egner, H., Schorch, M., & Voss, M. (Eds.), *Learning and Calamities: Practices, Interpretations, Patterns* (pp. 125–142). New York: Routledge.

Lidstone, J. (1996). Disaster Education: Where We Are and Where We Should Be. In Lidstone, J. (Ed.), *International Perspectives on Teaching about Hazards and Disasters* (pp. 7–18). Clevedon, UK: Channel View Publications.

Lübken, U. (2015). *Historia Magistra Vitae*, as the Saying Goes: Why Societies Do Not Necessarily Learn from Past Disasters. In Egner, H., Schorch, M., & Voss, M. (Eds.), *Learning and Calamities: Practices, Interpretations, Patterns* (pp. 112–122). New York: Routledge.

Manfredi, G., Rose, A., Sapountzaki, K., Jørgensen, G., Callaghan, E., Tobin, G., ... & Asprone, D. (2014). Resilience and Sustainability in Relation to Disasters: A Challenge for Future Cities: Common Vision and Recommendations. In *Resilience and Sustainability in Relation to Natural Disasters: A Challenge for Future Cities* (pp. 77–79). Cham: Springer.

Marincioni, F. (2007). Information technologies and the sharing of disaster knowledge: The critical role of professional culture. *Disasters, 31*(4), 459–476. doi:10.1111/j.1467-7717.2007.01019.x.

McGinn, N. F. (1985). A menu of possible educational responses in the context of disasters. *Disasters, 9*, 13–21. doi:10.1111/j.1467-7717.1985.tb00957.x.

Mercer, J., Kelman, I., Suchet-Pearson, S., & Lloyd, K. (2009). Integrating indigenous and scientific knowledge bases for disaster risk reduction in Papua New Guinea. *Geografiska Annaler. Series B, Human Geography, 91*(2), 157–183. doi:10.1111/j.1468-0467.2009.00312.x.

Mercer, J., Kelman, I., Taranis, L., & Suchet-Pearson, S. (2010). Framework for integrating indigenous and scientific knowledge for disaster risk reduction. *Disasters, 34*(1), 214–239. doi:10.1111/j.0361-3666.2009.01126.x.

Mudavanhu, C., Manyena, S. B., Collins, A. E., Bongo, P., Mavhura, E., & Manatsa, D. (2015). Taking children's voices in disaster risk reduction a step forward. *International Journal of Disaster Risk Science, 6*(3), 267–281. doi:10.1007/s13753-015-0060-7.

O'Hern, D. M., & Nozaki, Y. (2014). *Natural Science Education, Indigenous Knowledge, and Sustainable Development in Rural and Urban Schools in Kenya: toward Critical Postcolonial Curriculum Policies and Practices* (Vol. 6). Rotterdam: SensePublishers.

Okpala, J. (1996). The Treatment of Natural Hazards in Geography Textbooks for Senior Secondary Schools in Nigeria. In Lidstone, J. (Ed.), *International Perspectives on Teaching about Hazards and Disasters* (pp. 103–130). Clevedon, UK: Channel View Publications.

Petal, M. (2015). Critical Reflections on Disaster Prevention Education. In Egner, H., Schorch, M., & Voss, M. (Eds.), *Learning and Calamities: Practices, Interpretations, Patterns* (pp. 159–178). New York: Routledge.

Pfister, C. (2015). 'The Monster Swallows You': Disaster Memory and Risk Culture in Western Europe, 1500–2000. In Egner, H., Schorch, M., & Voss, M. (Eds.), *Learning and Calamities: Practices, Interpretations, Patterns* (pp. 77–93). New York: Routledge.

Quayson, A. (2000). *Postcolonialism: Theory, Practice or Process*. London: Polity.

Rattien, S. (1990). The role of the media in hazard mitigation and disaster management. *Disasters, 14*(1), 36–45. doi:10.1111/j.1467-7717.1990.tb00970.x.

Reiff, H. (1985). Educational development, basic human needs and disaster relief. Lessons from a few experiences. *Disasters, 9*, 5–12. doi:10.1111/j.1467-7717.1985.tb00956.x.

Rivera, J. D., & Miller, D. S. (2009). Disaster vulnerability education: A new focus on disaster education across the curriculum. *Journal of Applied Security Research, 4*(1–2), 60–67. doi:10.1080/19361610802210202.

Rodríguez, H., Quarantelli, E. L., & Dynes, R. R. (2007). *Handbook of Disaster Research*. New York: Springer.

Romo-Murphy, E., James, R., & Adams, M. (2011). Facilitating disaster preparedness through local radio broadcasting. *Disasters, 35*(4), 801–815. doi:10.1111/j.1467-7717.2011.01234.x.

Shaw, R., Mallick, F., & Takeuchi, Y. (2011). Essentials of higher education in disaster risk reduction: Prospects and challenges. In Shaw, R., Shiwaku, K., & Takeuchi, Y. (Eds.), *Disaster Education* (pp. 95–113). Bingley: Emerald Group Publishing Limited.

Shaw, R., Takeuchi, Y., & Shiwaku, K. (2011). '*Tsunagaru*': The Essence of Disaster Education. In Shaw, R., Shiwaku, K., & Takeuchi, Y. (Eds.), *Disaster Education* (pp. 153–162). Bingley: Emerald Group Publishing Limited.

Shiwaku, K., & Fernandez, G. (2011). Roles of School in Disaster Education. In Shaw, R., Shiwaku, K., & Takeuchi, Y. (Eds.), *Disaster Education* (pp. 45–75). Bingley: Emerald Group Publishing Limited.

Soffer, Y., Goldberg, A., Avisar-Shohat, G., Cohen, R., & Bar-Dayan, Y. (2010). The effect of different educational interventions on schoolchildren's knowledge of earthquake protective behaviour in Israel. *Disasters, 34*(1), 205–213. doi:10.1111/j.1467-7717.2009.01125.x.

Spiekermann, R., Kienberger, S., Norton, J., Briones, F., & Weichselgartner, J. (2015). The Disaster-Knowledge Matrix – Reframing and evaluating the knowledge challenges in disaster risk reduction. *International Journal of Disaster Risk Reduction, 13*, 96–108. doi:10.1016/j.ijdrr.2015.05.002.

Tait, N. (1996). Studying Natural Hazards in South African Schools. In Lidstone, J. (Ed.), *International Perspectives on Teaching about Hazards and Disasters* (pp. 59–70). Clevedon, UK: Channel View Publications.

Tikly, L., & Bond, T. (2013). Towards a postcolonial research ethics in comparative and international education. *Compare: A Journal of Comparative and International Education, 43*(4), 422–442.

Tobin, G. A., Whiteford, L. M., Murphy, A. D., Jones, E. C., & McCarty, C. (2014). Modeling Social Networks and Community Resilience in Chronic Disasters: Case Studies from Volcanic Areas in Ecuador and Mexico. In *Resilience and Sustainability in Relation to Natural Disasters: A Challenge for Future Cities* (pp. 13–24). Cham: Springer.

United Nations. (1994). *Yokohama Strategy and Plan of Action for a Safer World: Guidelines for Natural Disaster Prevention, Preparedness and Mitigation.* Yokohama, Japan: United Nations Retrieved from www.unisdr.org/eng/hfa/hfa.htm.

United Nations. (1999). *Proceedings: International Decade for Natural Disaster Reduction (IDNDR).* Geneva: United Nations. Retrieved from www.unisdr.org/files/31468_programmeforumproceedings.pdf.

United Nations. (2005). *Hyogo Framework for Action 2005–2015: Building the Resilience of Nations and Communities to Disasters.* Kobe, Japan: United Nations A/CONF.206/6. Retrieved from www.unisdr.org/eng/hfa/hfa.htm.

United Nations. (2015). *Sendai Framework for Disaster Risk Reduction 2015–2030.* Geneva: United Nations. Retrieved from www.unisdr.org/who-we-are/international-strategy-for-disaster-reduction.

United Nations. (n.d.). International strategy for disaster reduction. Retrieved from www.unisdr.org/who-we-are/international-strategy-for-disaster-reduction.

United Nations International Strategy for Disaster Reduction (UNISDR). (2009). *2009 UNISDR Terminology on Disaster Risk Reduction.* Geneva: UNISDR. Retrieved from www.unisdr.org/files/7817_UNISDRTerminologyEnglish.pdf.

van Voorst, R. (2016). Formal and informal flood governance in Jakarta, Indonesia. *Habitat International, 52*, 5–10.

van Voorst, R., Wisner, B., Hellman, J., & Nooteboom, G. (2015). Introduction to the 'risky everyday'. *Disaster Prevention and Management, 24*(4). doi:10.1108/DPM-04-2015-0077.

Walshe, R. A., & Nunn, P. D. (2012). Integration of indigenous knowledge and disaster risk reduction: A case study from Baie Martelli, Pentecost Island, Vanuatu. *International Journal of Disaster Risk Science, 3*(4), 185–194. doi:10.1007/s13753-012-0019-x.

Weichselgartner, J., & Pigeon, P. (2015). The role of knowledge in disaster risk reduction. *International Journal of Disaster Risk Science,* 6(2), 107–116. doi:10.1007/s13753-015-0052-7.

White, G. F., Kates, R. W., & Burton, I. (2001). Knowing better and losing even more: The use of knowledge in hazards management. *Global Environmental Change B: Environmental Hazards,* *3*(3), 81–92. doi:10.1016/S1464-2867(01)00021-3.

Whitehead, P. (1996). Natural Hazards Education: A Question of Implementation Strategies. In Lidstone, J. (Ed.), *International Perspectives on Teaching about Hazards and Disasters* (pp. 19–32). Clevedon, UK: Channel View Publications.

7 Curriculum development on climate change adaptation

Pre-service teacher training in Mongolia

Yembuu Batchuluun and Getsel Uranchimeg

Education for Sustainable Development (ESD) is an integral element of quality education and of all efforts to achieve sustainable development to support the integration of ESD in education policies, plans, curricula, pedagogy and assessment through evidence-based advocacy, technical assistance and monitoring, thus ensuring an effective follow-up to the United Nations Decade of ESD. Within this framework, climate change education (CCE) fosters understanding of the complexities and interconnection of the various challenges posed by climate change. CCE for sustainable development has a central role to play in helping the public, and especially the next generation, to understand and relate to the issues, make lifestyle changes to reduce greenhouse gas emissions and adapt to the changing local conditions. While CCE is needed at all levels, particularly during general education levels, instilling climate change awareness and understanding at a young age is ultimately the best way to change behaviours and attitudes. Education and awareness-raising play an essential role in increasing the climate change adaptation and mitigation capacities of communities by enabling individuals to make informed decisions. Education helps learners understand the causes and consequences of climate change, prepares them to live with the impacts of climate change and empowers women and men to adopt more sustainable lifestyles. As part of its work on Education for Sustainable Development, UNESCO supports countries to integrate climate change into their education systems' (UNESCO, 2014a, 2014b). Mongolia is one of the countries most vulnerable to climate change. Aridity in the country is not the only relevant factor to climate change impact. Low awareness of the climate change adaptation and mitigation is critically important to combat climate change in Mongolia. This chapter introduces a pilot study of curriculum implementation on 'Climate Change Adaptation' for pre-service teacher training, developed by the Geography Department, Mongolian State Pedagogical University of Education (MSUE).

The importance of the climate change adaptation (CCA) curriculum in Mongolia

The following reasons were identified for the importance of CCA curriculum development:

1. ***The global agendas and new paradigms of education***: By raising awareness and promoting knowledge and skills-development, education is an essential component and a catalyst for responding to global climate change. Its importance has been increasingly highlighted at the international level. In particular, Article 6 of the UN Framework Convention on Climate Change (UNFCCC) encourages parties to promote, develop and implement educational, training and public awareness programmes on climate change and its effects. In addition, the United Nations General Assembly proclaimed the UN Decade of Education for Sustainable Development (DESD) 2005–2014, emphasising that climate change is one of the key action themes of the decade. Article 2 of the Paris Agreement (2015) emphasised the need to strengthen responses to climate change by increasing the ability of communities to adapt. UNFCCC Article 6 (UNFCCC, 2015), climate change education for sustainable development (UNESCO, 2010) and Sustainable Development Goal 13 (United Nations, 2015) calls for the development of education and training programmes on climate change and its effects so as to have a well-informed and enlightened world citizenry.
2. ***The current environmental issues and climate change impacts in the country***: Mongolia is very sensitive to climate change due to its geographic location, sensitive ecosystems and socio-economic conditions (Dagvadorj, et al., 2009). About 90 per cent of the total territory belongs to arid and semi-arid zones, prone to desertification. Over a period of 30 years, 46.5 per cent of agricultural area suffered from wind and water erosion, and 70 per cent of pasturelands have been overgrazed (MNEGD, 2015). The Desertification Atlas of Mongolia (2013) showed that 77.8 per cent of total Mongolian territory was affected by desertification to differing degrees (Bulgan et al., 2013; MNEGD, 2013). In comparison to 2006, the statistics prove that the percentage of Mongolian land that is vulnerable has now risen by 2 to 3 per cent. Land degradation induced by anthropogenic factors is common to all ecological zones of Mongolia. There are several specific human activities that have led to serious and widespread soil erosion and land degradation. Two factors are the primary causes of desertification, and 40 per cent of Mongolian territory has been affected by desertification because of natural causes, while the remaining 60 per cent of desertification is the result of harmful human activities. Mining and inadequate waste management can also

make significant contributions to land degradation. In addition, the accelerated rate of economic development and infrastructure construction has impacted the nomadic lifestyle and natural ecosystems, and traditional pastoral livestock production as well as traditional knowledge, local wisdom which was more sustainable before (Yembuu, 2016).

3 ***Lack of public awareness and educational institutions' role in climate change in Mongolia***: Education plays a paramount role in raising awareness and promoting behavioural change for climate change mitigation and adaptation. Educators and trainers are powerful agents of change for delivering the educational response to sustainable development, but for them to help usher in the transition to a sustainable society, they must first acquire the necessary knowledge, skills, attitudes and values (UNESCO, 2014c). Even though climate change is a big challenge in Mongolia, many people do not know what it is, or how it affects their lives. So to raise public awareness, knowledge and attitudes is relatively more important than to recover the degraded land and environment. CCA reduces the impacts of climate change on human livelihoods and builds resilience while exploiting beneficial opportunities. Educational institutions' roles are very important in adapting to climate change.

According to UNESCO, if pre-service and in-service teachers learn to weave ESD issues into the curriculum and to use pedagogical techniques associated with quality ESD, then the next generation will be capable of shaping a more sustainable world (UNESCO, 2005). But the lack of teacher knowledge and understanding remains the primary barrier to the effective implementation of child and learner-centred sustainable development curricula in the schools. The fifth objective of the National Action Program in Climate Change (was redeveloped and approved in 2011) was 'to conduct public awareness campaigns and support citizen and community participation in actions against climate change', including courses about climate change and the green economy in all levels of school programmes; develop educational curricula related to climate change and environmental sectors. In reality, implementation of this objective is still not sufficient, is limited by the 'show-based' events of 'Water day', 'Earth day' and 'Planting Tree Days', etc in Mongolia (Navchaa & Batchuluun, 2012).

Simplified resources and systematical training are needed for teachers and the younger generation: there are many paper-based materials, reports, research works and international projects implemented in Mongolia on climate change issues. However, they are mostly not relevant to the real needs of users, but are only 'unclear measurement' orders such as 'to decrease the air pollution; to reduce coal consumption; to increase energy efficiency' etc. with no clear answers to 'HOW' these should be done. Somehow, scientific research and model-based trends of air temperatures, precipitations bring 'awfully-oriented' information to the public. Many respondents answered that 'I can do nothing for against the climate change' in a recent survey (Batchuluun, Uranchimeg, & Enkhjargal, 2015). Therefore, simplified information on climate change is needed for the public and the curriculum must be clear with participation-oriented measures on climate change adaptation. CCE should be a life-long process starting at an early age and continuing through all learning stages so as to promote behavioural change and enhance the resilience of communities.

Methodology of needs assessment survey

The content and methodology of the curriculum based on the needs assessment survey and research study consists of three main analyses: (1) Review of policy; (2) Review of national educational standards, curricula, textbooks, and (3) Review of teachers' knowledge of climate change. A policy analysis was conducted via a matrix (PAM) relating to CCE in Mongolia. PAM comprised of two sections: the first section reviewed 37 educational documents, such as educational laws, national programmes adopted by the orders of the minister for education and science. The second section covered 12 laws, 16 national programmes and other policy documents relating to environmental management, climate change and disaster risk reduction, such as Law on Preventing Disasters, Law on Protected Areas, National Program on Climate Change, National Program on Water Resources, National Program on Coping with Desertification, Law on Education; 25 subjects' standards, curricula and textbooks were also reviewed in the baseline survey.

Quantitative/qualitative analysis defined the level of teacher knowledge and the needs of the training on CCE for sustainable development resulted in a 25-question survey being developed. In addition to the Likert scale-based questionnaire the qualitative component was made via semi-structured interviews with research participants. This questionnaire was applied to approximately 150 teachers in urban and rural areas. A total of 94 per cent of respondents were women, and 6 per cent were men. The dominance of female teachers in our sample matches the overall ratio of male and female teachers in the country. Over 80 per cent of secondary school teachers in Mongolia are female. Rural teachers were a target for focus group discussions.

The following are example questions of focus group discussions:

- Have you seen any empty classrooms with the lights left on?
- Is there any faucet in your school toilet that leaks?
- Do teachers use both sides of the paper when printing or making photocopies?

- Do you have garbage bins at your school for sorting garbage?
- What are the main environmental issues that you are facing at the local level? Are there ways to resolve them?
- How much do parents participate in renovating classrooms and gardening school areas?
- What aspects of indigenous knowledge have been forgotten in your area?
- Have you attended any training on CCE?

A total of 46 per cent of the respondents answered that the questions in the questionnaire related to their course and 39 per cent think that these issues are deeply involved while only 5 per cent answered as related and necessary to be taught at schools (Navchaa, Batchuluun, Amarbaysgalan, & Uranchimeg, 2016).

Although most of the respondents agree that concepts relate to their courses, teachers have different thoughts on each of the questions depending on their professional background. For example, primary school, Mongolian and foreign language teachers think that teaching about cultural differences and the misunderstandings associated with it and maintenance of traditional culture is deeply related, while natural and social science teachers answer that issues being faced by people – climate change, social and human ecological issues and changes in the relationship between people and nature – would be the deeply related ones. The study shows that subjects related to climate change concepts besides geography mostly focus on academic knowledge about the climate, and are not relevant to an individual change in the lifestyle of the younger generation (Batchuluun et al., 2015).

Curriculum content development

As a response to the needs assessment survey, a new curriculum development approach is proposed. The stages are shown in Figure 7.1. The needs assessment study shows that CCE should not be restricted to pure scientific or factual knowledge, but should develop and encourage an organic awareness of the importance of everyday sustainable practices (such as recycling, not leaving the devices turned on, and printing on both sides of paper). Therefore, the CCA curriculum was developed to focus on daily needs, lifestyle changing-oriented activities and aimed to promote the knowledge, skills, attitudes of climate change and values necessary to shape a sustainable future. The CCA curriculum was structured on three modules (1) Climate Change and Ecosystem, (2) Climate Change and Society, (3) Ecosystem-based Adaptation and Mitigation.

Knowledge, skills, attitudes and assessment criteria of learners are described in each theme of the modules. Each theme consists of an objective, learning methodologies and steps for activities which are based

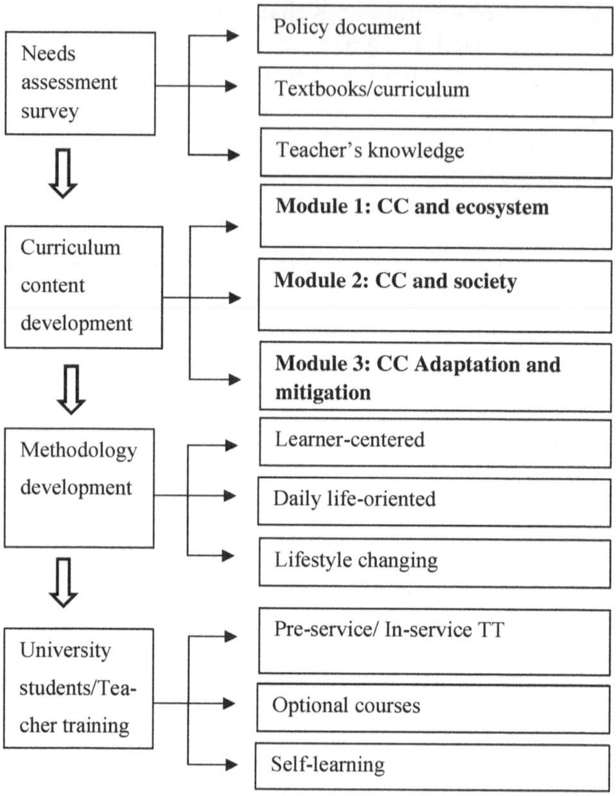

Figure 7.1 Curriculum Development Stages and Structure on Climate Change Adaptation.

on daily life cases, critical thinking questions, self-learning oriented investigations as well as scientific knowledge. The learner-centred teaching/ learning methodologies of the curriculum are more focused on inquiry-based and project-based learning as well as problem-solving skills (Figure 7.2). An example of a critical thinking activity in the CCA curriculum is as follows.

The example provided above is used for teacher education and other stakeholders as a manual or guideline of climate change adaptation issues. Over the past two years, the curriculum on climate change adaptation developed by MSUE is offered as an optional course for pre-service and in-service teacher training by the Geography Department of MSUE. Early observations show that this curriculum and other connected activities with climate change education are influencing the students' behavioural change, such as saving water and paper. Although it is a small step forward, adaptation should start from the simple needs and lifestyles.

Am I an energy-friendly person?

Discuss how saving energy will mitigate desertification.

How much energy can we save? How many light bulbs do you have in your home? For example: a lamp, kitchen light, bathroom etc. How many of them are incandescent light bulbs, and how many are not? It is estimated that if you use a non-incandescent light bulb over an incandescent one, you will save 68.1 kg of carbon dioxide per year. If you change all the light bulbs to non-incandescent bulbs, how much carbon dioxide will you save? Using a non-incandescent light bulb will save your energy consumption by 50%. How much can you save on your monthly electricity bill? How much can you save this year? *Discuss the findings*: Can you mitigate desertification by switching off unused lights and changing to non-incandescent light bulbs? Please raise the question on topic "Am I an energy saver?" and discuss it in your class.

Figure 7.2 Critical thinking activity example in the CCA curriculum

Conclusion

In the context of Mongolia, analysis of the school curriculum shows that climate-related issues are only taught in Human and Nature subjects at primary level and in geography and science subjects at secondary level. In tertiary institutions, climate change issues are mostly taught in Natural Sciences. Consequently, students who do not take these subjects/courses, lack the basic understanding and skills relevant to climate change adaptation. Therefore, climate change curricula at all levels should be inclusive, not discriminatory.

There is an abundance of legal and policy documents related to the climate change issues in Mongolia, but there has been limited coverage of climate change education and unclear implementation of the documents. Activities to improve public understanding of CCE policies, and clear expected outcomes should be added to national programmes. The legal documents should clearly say what should be done under the current capacity-building programmes for climate change education for sustainable development.

Teacher knowledge and skills have been limited in teaching climate change adaptation. Cross-curricular activities are very limited in the national curriculum of education. Teachers' methodology is still theoretically-oriented and teacher-centred. Therefore, teachers need up to date teaching aids and practices and create student-centred learning to help them efficiently transfer their knowledge to students. A more sustainable society requires climate change adaptation, informed citizens and the knowledge and the skills to contribute towards those goals of sustainability. This needs to be enabled by a more informed teaching workforce.

References

Batchuluun, Y. (2016). Chapter 7. Mongolian Nomads: Effects of Globalization and Social Change. In Robertson, M., & Tsang, P. K. E. (Eds.), *Everyday Knowledge, Education and Sustainable Futures, Education in the Asia-Pacific Region: Issues* (p. 94). Singapore: Springer Science+Business Media.

Batchuluun, Y., Uranchimeg, G., & Enkhjargal, P. (2015). Contribution of Geography for Green Development in Mongolia. In Nyamdavaa, G. (Ed.), *Natural Resource and Territorial Condition for Socia-economic Development of Mongolia* (pp. 82–86). Ulaanbaatar. MUN Press.

Bulgan, D., Mandakh, N., Odbayar, M., Otgontugs, M., Tsogtbaatar, J., Elbegjargal, N., ... & Erdenetuya, M. (2013). *Desertification Atlas of Mongolia*. Ulaanbaatar, Mongolia: Institute of Geoecology, Mongolian Academy of Sciences and Environmental Information Centre, Ministry of Green Development.

Dagvadorj, D., Natsagdorj, L., Dorjpurev, J., & Namkhainyam, B. (2009). *Mongolia: Assessment Report on Climate Change 2009*. Ulaanbaatar: MNET.

Ministry of Nature, Environment and Green Development, Mongolia (MNEGD). (2013). *Desertification Atlas of Mongolia* (Edited by Dash, D. et al). Ulaanbaatar. Admon Publishing.

Ministry of Nature, Environment and Green Development, Mongolia (MNEGD). (2015). *Environmental Impact Assessment of Mongolia* (p. 57). Ulaanbaatar. Admon Publishing.

Navchaa, T., & Batchuluun Y. (2012). *A Public Awareness Programme of Ecosystem-Based Adaptation Principles and Practice*. Ulaanbaatar: Ministry of Environment and Tourism of Mongolia. doi:10.13140/RG.2.2.30021.45281.

Navchaa, T., Batchuluun, Y., Amarbaysgalan, G., & Uranchimeg, G. (2016). Policy provisions and teachers' needs on climate change education for sustainable development in Mongolia. *International Journal of Asian Business and Information Management,* 7(4), October–December, 36–48.

Paris Agreement. (2015). United nations framework convention on climate change – Paris Agreement. Paris, France. Retrieved from https://unfccc.int/files/meetings/paris_nov_2015/application/pdf/paris_agreement_english_.pdf.

United Nations. (2015). Sustainable development goals. Retrieved January 2017 from www.undp.org/content/undp/en/home/sustainable-development-goals.html.

United Nations Educational, Scientific and Cultural Organization (UNESCO). (2010). *Climate Change Education for Sustainable Development*. Paris. UNESCO.

United Nations Educational, Scientific and Cultural Organization (UNESCO). (2014a). 37 C/4. *Medium-Term Strategy (2014–2021)*. Paris: UNESCO. Retrieved March 2017 from http://unesdoc.unesco.org/images/0022/002278/227860e.pdf.

United Nations Educational, Scientific and Cultural Organization (UNESCO). (2014b). *UNESCO Education Strategy 2014–2021*. Paris: UNESCO. Retrieved

March 2017 from http://unesdoc.unesco.org/images/0023/002312/231288e.pdf.

United Nations Educational, Scientific and Cultural Organization (UNESCO). (2014c). *UNESCO Roadmap for Implementing the Global Action Programme on Education for Sustainable Development.* Paris: UNESCO. Retrieved March 2017 from http://unesdoc.unesco.org/images/0023/002305/230514e.pdf.

8 Fieldwork as a vehicle for sustainability education

The centrality of geographical inquiry

Niranjan Casinader and Gillian Kidman

The Sustainability Curriculum Framework and the Australian Curriculum

The curriculum currently in use in Australia pertaining to sustainability was first implemented in 2011, following the writing of the current national curriculum documents between 2009 and 2011. To understand the nature of this sustainability education curriculum, we need to explore the origins of the ideals being promoted at the time, as these provided the impetus for the method of inclusion of sustainability as a compulsory part of the national curriculum.

The years prior to 2009 saw strong support for a considered curriculum approach to sustainability education. Skamp (2010) conducted a 20-year review (1990–2009) of school practices and research of Education for Sustainability from kindergarten to Year 12, writing of the need for a process and participatory form of sustainability education. He advocated for a teaching *as* sustainability that would have a 'transforming purpose' on the student, as ownership of the learning is an essential factor in building one's ability to be an agent for change: learning needs to be 'meaningful, engaging and participative, rather than functional, passive and prescriptive' (Skamp, 2010, pp. 26–27). Payne and Wattchow (2009) described the importance of specific pedagogies that relate to experiences away from the classroom:

> There needs to be a shift in emphasis from focusing primarily on the 'learning mind' to re-engaging the active, perceiving, and sensuous corporeality of the body with other bodies ... in making-meaning in, about, and for the various environments and places in which those bodies interact and relate.
>
> (p. 16)

The *Earth Citizenship: Background Paper for Learning for Sustainability* (2009) provided a comprehensive meta-review of the research literature and identified the knowledge and practices required to create more sustainable futures. This document provided the basis for the Department of the Environment, Water, Heritage and the Arts to produce the *Sustainability Curriculum Framework* (2010) for Australia's national, state and territory curriculum developers and policy makers. This framework supports a progression of learning from kindergarten to Year 10, the core years of Australian education. This document advocates that an Australian sustainability education should be both for the present *and* for the future. An emphasis was made that our youth need to be able to 'design and implement actions for the present, in the knowledge that the impact of these actions will be experienced in the future' (Department of the Environment, Water, Heritage and the Arts, 2010, p. 4). Students are empowered to act, individually and collectively, as they explore and evaluate issues, as they gather evidence and create solutions for their sustainable future. This implies that students actively participate in sustainability issues, develop skills and learn to take responsibility in relation to issues they perceive to be of importance.

An inquiry-like action process, called the *sustainability action process* is the central organiser in the framework to highlight 'that education for sustainability is not simply the acquisition of knowledge or skills, but a total approach which generates motivation and commitment to take sustainability action for improved outcomes for a sustainable world' (Department of the Environment, Water, Heritage and the Arts, 2010, p. 8). The framework advocates that Education for Sustainability needs to be cross-disciplinary, as no single school subject or year level provides all the knowledge and skills; the knowledge and skills need to be interconnected across the disciplines, and over time. However, a cross-disciplinary approach, as recommended, is a challenge to curriculum developers and policy makers, as it represents non-traditional ways in the imagining of the knowledge, skills and dispositions inherent within a curriculum. Critically important is the interdependent and overlapping nature of the framework's key elements: sustainability action process, knowledge of ecological and human systems, and repertoires of practice.

The 2011 release of the Australian Curriculum was a staged process, beginning with the learning areas (school subjects) of English, Mathematics, Science and History. This release included sustainability as both a concept and as a cross-curriculum priority, as advocated in the framework of the Australian Curriculum. In science, the Year 4 child is introduced to the concept of sustainability through the study of the Earth's surface changing over time as a result of natural processes and human activity. The idea of human impact is consolidated in Year 7 where the interactions between organisms, including the effects of human activities, can be represented by food chains and food webs. Year 9 and 10 science emphasise the interdependent nature of organisms and abiotic components. However, the wording of the science curriculum implies a teaching of facts, teaching learners what to think and employing learning outcomes that reinforce the dominant social paradigm of Western society; for example the Earth's surface changes over time as a result of natural processes and human activity (ACSSU075) accompanied by the elaboration: 'Considering how different human activities cause erosion of the Earth's surface.'

In 2013, geography was released as a component of the Humanities curriculum. Geography's reaction to sustainability as a concept is far more extensive than that of science. Whereas science published 5 content descriptors in 4 year levels (4, 7, 9 and 10) relating to sustainability, geography published 24 content descriptors in all year levels from 1 to 10. In science, the sustainability concepts are all contained in the *Science Understanding* strand, indicating they are content knowledge. Geography also houses the sustainability concepts in its *Knowledge and Understanding* strand throughout both the primary and secondary years. Year 4 in geography has an environmental sustainability emphasis examining how the environment supports humans, the meaning of sustainability and how we need to apply the concept of sustainability to the management of resources and waste.

However, in the secondary school, Years 8, 9 and 10 also have sustainability housed in its *Geographical Inquiry Skills* strand. For example, in Year 10, the final year of compulsory school attendance in Australia, the unit of study is *Environmental change and management* which investigates environmental geography through an in-depth study of a specific environment in Australia and one other country. The unit considers the environmental functions that support all life, the major challenges to their sustainability, and the environmental world views that influence how we perceive and respond to these challenges. Human-environment systems thinking is applied to understand the causes and consequences of environmental change. Given this in-depth comparative case study emphasis, it is not surprising to find the *Geographical Inquiry Skills* strand mandates a participatory approach:

> Reflect on and evaluate findings of an inquiry to propose individual and collective action in response to a contemporary geographical challenge, taking account of environmental, economic, political and social considerations; and explain the predicted outcomes and consequences of their proposal.
>
> (ACHGS080)

The language used is not as confining as that used in the science curriculum. The language use in the Geography learning area implies a transformative building of knowledge open to all values, views and goals and is open to differing interpretations:

The elaborations that accompany this same content descriptor are:

- reflecting on the role of personal values and attitudes in influencing their responses to situations including goals (for example, environmental protection)

- explaining how the application of geographical concepts and methods has contributed to a deep understanding of the causes of and solutions to issues related to environmental change, human well-being or development

Although not explicitly stated, fieldwork would be an ideal context for this in-depth study. A fieldwork inquiry activity would allow for the participatory engagement advocated in sustainability education. The students would have the opportunity to be active outdoors, to safely explore their surroundings and to make connections between care and use of their environment (Reynolds, 2014). This is a form of a place-responsive pedagogy as it involves explicit efforts to teach *in* an environment *of* an environment, with the aim of understanding and improving that environment and its human-environment relations (Mannion, Fenwick, & Lynch, 2013). The fieldwork associated with the sustainability content of the geography curriculum would therefore be grounded in experimental and cooperative learning, involving the clarification of student values and a critical reflection that would enable the envisaging of a sustainable future. This interactive value-laden nature of subject matter is highly applicable to sustainability education, more so than the didactic nature of the scientific approach. These ideas will be explored further in the fieldwork section of this chapter. But before we explore the nature of fieldwork to develop sustainability knowledge and understandings, we first need to examine how sustainability is considered in the curriculum, what key concepts of sustainability knowledge are mandated, and the nature of the sustainability understandings that are being sought.

In the Australian Curriculum, sustainability is also incorporated as one of three mandatory cross-curriculum priorities that collectively encourage complementary learning and teaching across different disciplines. This is a response to the Sustainability Curriculum Framework recommendation for a cross-curricular approach to sustainability education. As a cross-curriculum priority, sustainability is given particular salience to ensure that it is attended to throughout the curriculum by teachers. The role of the cross-curriculum priorities is elevated in the development of curricula, instruction and assessments. It is intended, therefore, that sustainability should become a common and familiar concept across the disciplines and year levels. 'It should explicitly emerge in these multiple disciplinary contexts, thus enabling the student to develop a cumulative, coherent, and usable understanding of the priority' (Casinader & Kidman, 2018, p. 6).

The sustainability cross-curriculum priority is presented as three key concepts and nine organising ideas as shown in Table 8.1. From Table 8.1, we can see that sustainability includes a consideration of the interdependence of the environmental spheres (***systems***): (the interdependent and dynamic nature of systems that support all life on Earth and our collective well-being (the biosphere, atmosphere, hydrosphere and lithosphere)).

Table 8.1 Sustainability cross-curriculum priority key concepts and organising ideas (OI)

Systems	
OI.1	The biosphere is a dynamic system providing conditions that sustain life on Earth.
OI.2	All life forms, including human life, are connected through ecosystems on which they depend for their wellbeing and survival.
OI.3	Sustainable patterns of living rely on the interdependence of healthy social, economic and ecological systems.
World views	
OI.4	World views that recognise the dependence of living things on healthy ecosystems, and value diversity and social justice, are essential for achieving sustainability.
OI.5	World views are formed by experiences at personal, local, national and global levels, and are linked to individual and community actions for sustainability.
Futures	
OI.6	The sustainability of ecological, social and economic systems is achieved through informed individual and community action that values local and global equity and fairness across generations into the future.
OI.7	Actions for a more sustainable future reflect values of care, respect and responsibility, and require us to explore and understand environments.
OI.8	Designing action for sustainability requires an evaluation of past practices, the assessment of scientific and technological developments, and balanced judgements based on projected future economic, social and environmental impacts.
OI.9	Sustainable futures result from actions designed to preserve and/or restore the quality and uniqueness of environments.

Note
* Sourced from the Australian Curriculum (www.australiancurriculum. edu.au/crosscurriculum priorities/ sustainability/key-ideas).

Sustainability also includes a ***World View*** and ***Futures*** view. The ***World View*** organising ideas focus on the relationships between living things (including people) and their natural environment. The ***Futures*** component highlights long-term viability as central to sustainability of not just the natural environment, but it also incorporates 'ecological, social and economic systems' (ACARA, ND). Teachers need to go beyond environmental education's focus on the systems view, and therefore only adopt a scientific inquiry approach. Such an approach is not addressing the full scope of sustainability education as defined by the Australian Curriculum. The

Futures concept aims to build capacities for thinking and acting in ways that are necessary for a more sustainable future. The concept involves reflective thinking processes that lead to a more equitable and sustainable future.

Sustainability is included in each learning area, but each of these contributes in a different way to the sustainability cross-curriculum priority, its key concepts and organising ideas. For example, science has content that enables students to work with ecological and human systems and to appreciate their interdependence. Geography contributes to the development of world views that encourage students to act to create a more socially and ecologically just world. It also provides content that challenges students to consider a sustainable future and to take action that recognises projected future economic, social and environmental impacts.

In summary, the content of Australia's sustainability curriculum, as a cross-curriculum priority, is based on a sound base of extensive reviews of research and programmes. However, it is our contention that, when teaching sustainability, from a content approach akin to that required by the *Australian Curriculum: Science*, the teacher is simply teaching sustainability as discipline content. In contrast, if the teacher taught sustainability through an inquiry process, then the teacher's approach becomes sustainability education, a transformative process.

The following section considers the implied fieldwork link with sustainability that we argue is necessary for sustainability education as a transformative process.

Defining sustainability education fieldwork and its benefits

In our earlier work, we have introduced the notion of a ***fieldwork imperative*** (Casinader & Kidman, 2017, 2018), where we distinguish between *why we do* fieldwork and *why we should do* fieldwork in relation to sustainability education. We use the phrase to highlight an existing bias of sustainability education from the science education base. The term implies the enormity of the 'necessity' of sustainability education, as a 'moral obligation' that includes an environmental perspective in the 'promise' of educating to preserve the planet. We defined the term *fieldwork imperative* to be 'the felt drive to use fieldwork to gain various forms of knowledge for its own sake, as well as a motive to achieve a worthy practical end' (Casinader & Kidman, 2017). To accompany this definition, we now feel the need to clarify the term *fieldwork* as included in this definition, given that we wish to reconceptualise fieldwork and distinguish between *why we do* fieldwork and *why we should do* fieldwork. Consider the following definition of fieldwork provided to Australian teachers by ACARA (the Australian Curriculum authority): 'Any activity involving observation and recording of information outside a classroom. It could be within the school grounds, around neighbouring areas or in more distant locations' (ACARA, n.d.). This definition is, we feel, limiting for sustainable education purposes, as it focuses on the *why we do* fieldwork – to observe and record information (and on the *where* – possible locations). It does not inform the teacher on the *why we should do fieldwork*: that it can generate new knowledge for the student as well as promote their personal growth, both in the now as well as in the future. The teacher is not informed that fieldwork has its own *internal drive*, that of learning for *now* and for the *future*.

To further develop this conceptual context of fieldwork, the question we now raise is: what is this *internal drive* of fieldwork? We draw upon the words of Callahan (2003), who describes a 'research imperative' from the health sciences in terms of a continuum. The Callahan continuum is bounded by the pursuit of knowledge for worthy goals, but with the acknowledgement and recognition of possible compromising risks. Callahan's continuum can be extrapolated to create the *internal drive continuum of sustainability education fieldwork*. This is shown in Figure 8.1.

In Figure 8.1, the continuum is represented as consisting of three points or stages: (1) the fieldwork imperative as the **drive to gain knowledge** for its own sake (for example, to explore the water quality of a local water course); (2) the fieldwork imperative as **an internalised moral obligation** to solve a problem (e.g. to identify the source of a water pollutant), and (3) the fieldwork imperative as a motivation to pursue **the goal of prevention** of a pollutant (e.g. to challenge the polluting agents with the goal of enforcing a change in behaviour). We consider these three elements to provide the meaning for the internal drive – the fieldwork imperative of *sustainability education fieldwork*. This internal drive continuum is the underlying concept of *why we should do* fieldwork, and is central to the thesis of this chapter.

During a fieldwork inquiry, the teacher needs to facilitate the students' movement along this continuum. It is important to state here that the continuum is not a simple linear binary, moving from acceptable at one end to not acceptable at the other; all fieldwork learning is worthy of value, however small that may be. However, we contend that *sustainability education fieldwork* goes beyond the collection of data to solve a problem. Fieldwork needs to go beyond the practical as the desire to collect 'facts and data' and form a conclusion. It often highlights a problem that is within the realm of student ability to determine a solution. As a result, there is a natural progression beyond the 'Conclusion' heading of a fieldwork or inquiry report to take practical action to improve the situation;

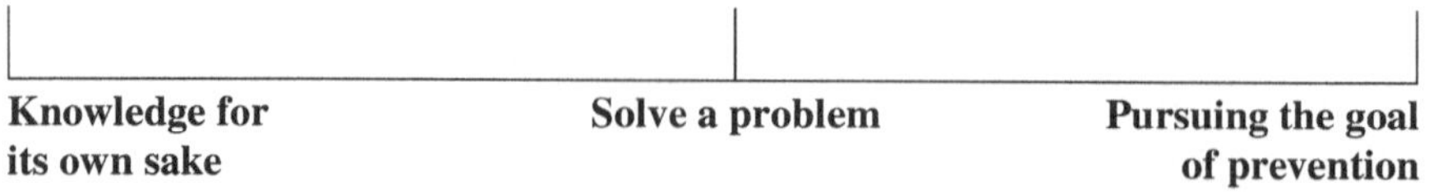

Figure 8.1 The internal drive of sustainability education fieldwork.

that is, for the student to act upon the short or long-term implications of the inquiry.

Fieldwork is necessary for knowledge generation for a sustainable future. We can no longer justify fieldwork that collects information for information's sake. Careful consideration needs to go into selecting what information needs to be collected, why it needs to be collected, how we are going to collect it, and importantly what we are going to do with the information and why. Thus, we need to consider the role that the information plays in the fieldwork. The need of working towards a preventative goal is a critical focus of *sustainability education fieldwork*.

In summary, we assert that fieldwork is an imperative for sustainability education. *Sustainability education fieldwork* has the particular characteristic of an internal drive for the pursuit to undertake fieldwork both now and in the future, for the future. This *internal drive* is in the form of a continuum that involves the collection and creation of knowledge not only for problem solving, but with the *belief of prevention* as an essential outcome.

Having established that there is a fieldwork imperative, and what the nature of that imperative is in terms of defining *sustainability education fieldwork*, we now query the impact of *sustainability education fieldwork* on young people in our society. By considering the limited research literature available, we summarise our tentative answers under three themes:

1 *Inquiry-based Skills:* Both observational and analytical skills are developed so that a habit of being mindful of immediate surroundings, especially familiar surroundings (Bliss, 2009), is developed. Additionally, learners are able to visualise, articulate, conceptualise, and solve problems and make informed sustainable decisions (Casinader & Kidman, 2018), which are necessary skills for a sustainable future;
2 *Affective Skills*: By experiencing how it feels to explore real-world situations through 'doing', and to create projects in or to simulate real-world conditions (Kleeman, 2009) develops a respect for the environment and an awareness of the impact of our interactions with nature and the environment (Bliss, 2009). The tools a student uses to interact with others socially in the learning space (Gold et al., 1991) strengthen the learning experience; and
3 *Personal Skills:* A sense of responsibility for their own learning is developed. This includes the idea that personal circumstances and other people are not responsible for the choices an individual makes in relation to their learning in- and out-of-school. The student grows from making mistakes (Perkins, Tishman, Ritchhart, Donis & Andrade, 2000), they also grow in terms of recognising their individual strengths, abilities and attributes (Gold et al., 1991).

Conclusion

Globally, schools need a sustainability curriculum that is 'holistic, multi-disciplinary and contextually relevant' (Holdsworth, Thomas & Hegarty, 2013, p. 355). Curriculum needs to go beyond simply the teaching of facts and reflecting the priorities of today's society. We now need to teach children to think critically about the nature of knowledge and to then use their knowledge to prepare for a sustainable future. As educators, we need to be aware of disciplinary biases and shortcomings. We need to be aware that we are all active learners and are all required to construct our own understandings and knowledge (Holdsworth et al., 2013).

In this context, it is our opinion that the Australian Curriculum does not provide a clear pathway for its teachers to address sustainability in the classroom. The curriculum provides sustainability concepts as disciplinary content in science that are simply reinforcing the acquisition of facts and Western science biases. In geography, the curriculum also provides sustainability concepts as disciplinary content. However, this content incorporates values, attitudes and an inquiry-based approach to solving issues of personal interest to the student. Further, the Australian Curriculum states that as a cross-curriculum priority, its content needs to be addressed through all the prescribed learning areas. Thus, there appear to be three components to sustainability that the Australian teacher needs to consider: a limiting Western science approach; the broad values approach of geography; and the cross-disciplinary priority of sustainability.

Since the pedagogical approaches in sustainability education that follow the didactic science approach revolve around an understanding of the mutual dependence between the four spheres of the natural environment, there may be a tendency to adopt this scientific approach at the expense of other disciplinary approaches. Consequently, any field-based scientific inquiry could be narrowly focused, and be more concerned with the systematic, precise and measured study of a particular site, in which the gathering of information is controlled and managed at every stage (Nielsen, Harbsmeier, & Ries, 2012). We are not implying that the natural systems components of sustainability should be ignored, but we do believe that it is necessary to point out that the systems approach tends to displace other inherent facets within the sustainability concept. For example, any discussion as to the sustainable use of our natural environments should incorporate discussions of the way people interact with these environments, and also include a personal exploration of an issue of interest to incorporate a values component to sustainability, but this is not the case in the Australian Curriculum.

In advocating a fieldwork imperative for sustainability education, we acknowledge that there are many likely instances worldwide where fieldwork is being used for sustainability education. To date, however, we have not found evidence of a specific fieldwork pedagogy that incorporates the internal drive to undertake fieldwork, both for the now and in the future, for the future. We conceptualise this internal drive to involve the collection and creation of knowledge, not only for problem solving, but with the belief of prevention as an essential outcome. We assert that there is an imperative for sustainability education to employ a *sustainability education fieldwork* recommendation that has the particular characteristic of an internal drive. Such an entity within the sustainability literature and in curriculum frameworks would assist in the development of: (1) inquiry-based skills that develop a habit of being mindful of one's surroundings; (2) affective skills of exploring the world by doing, and (3) the personal skills and a sense of responsibility.

There is ample evidence in the literature that asserts that simply making additions or changes to the curriculum will not ensure its effective delivery in the classroom. The development and inclusion of a *sustainability education fieldwork* imperative is indeed advocating a shift in educational practice. Research needs to be undertaken into the areas of appropriate disciplinary pedagogies, teaching and learning methods in order to determine the disciplinary approaches for *sustainability education fieldwork*. The time is long overdue for professional development programmes that have a dual function: working to systematically change school and personal cultures in order to develop a sustainability literacy; and enabling a shift away from existing fieldwork biases that may exist in a school.

References

Australian Curriculum, Assessment and Reporting Authority (ACARA). (n.d.). Sustainability. Retrieved 21 August 2017 from www.australiancurriculum.edu.au/f-10-curriculum/cross-curriculum-priorities/sustainability/.

Bliss, S. (2009). Fieldwork: The heart of geography. *Geography Bulletin, 41*, 7–11.

Callahan, D. (2003). *What Price Better Health? Hazards of the Research Imperative.* Berkeley: University of California Press.

Casinader, N., & Kidman, G. (2017). Fieldwork as a vehicle for Sustainability Education: The centrality of geographical inquiry. Conference paper presented at 5th Education for Sustainability Asia Conference, 2–3 May, National Institute of Education, Nanyang Technological University, Singapore.

Casinader, N. & Kidman, G. (2018). Fieldwork, sustainability and environmental education: The centrality of geographical inquiry. *Australian Journal of Environmental Education, 34*(1), 1–17.

Department of the Environment, Water, Heritage and the Arts. (2010). *Sustainability Curriculum Framework: A Guide for Curriculum Developers and Policy Makers.* Sydney: Commonwealth of Australia. Retrieved from https://cpl.asn.au/sites/default/files/journal/Sustainability%20Curriculum-Framework.pdf.

Gold, J. R., Jenkins, A., Lee, R., Monk, J., Riley, J., Shepherd, I. D. H., & Unwin, D. J. (Eds.). (1991). *Teaching Geography in Higher Education: A Manual of Good Practice.* Oxford, UK: Blackwell.

Holdsworth, S., Thomas, I., & Hegarty, K. (2013). Sustainability Education: Theory and Practice. In Dillon, J., Brody, M., Stevenson, R., & Wals, A. (Eds.),

International Handbook of Research on Environmental Education. New York: Routledge.

Kidman, G., & Casinader, N. (2017). *Inquiry-based Teaching and Learning across Disciplines: Comparative Theory and Practice in Schools*. Basingstoke, UK: Palgrave Macmillan.

Kleeman, G. (2009). Doing fieldwork with year 7. *Geography Bulletin, 41*, 2–6.

Mannion, G., Fenwick, A., & Lynch, J. (2013). Place-responsive pedagogy: Learning from teachers' experiences of excursions in nature, *Environmental Education Research, 19*(6), 792–809.

Nielsen, K. H., Harbsmeier, M., & Ries, C. J. (2012). Studying Scientists and Scholars in the Field: An Introduction. In Nielsen, K. H., Harbsmeier, M., & Ries, C. J. (Eds.), *Studying Scientists and Scholars in the Field: Studies in the History of Fieldwork and Expeditions*. Arhus/Kobenhavn, Denmark: Aarhus University Press.

Payne, P. G., & Wattchow, B. (2009). Phenomenological deconstruction, slow pedagogy and the corporeal turn in wild environmental/outdoor education. *Canadian Journal of Environmental Education, 14*, 15–32.

Perkins, D., Tishman, S., Ritchhart, R., Donis, K., & Andrade, A. (2000). Intelligence in the wild: A dispositional view of intellectual traits. *Educational Psychology Review, 12*(3), 269–293.

Reynolds, R. (2014). *Teaching Humanities and Social Sciences in the Primary School* (3rd ed.). Sydney: Oxford University Press.

Skamp, K. (2010). *Critical Review of Current Practice and Research of Environmental Education and Education for Sustainability for Kindergarten to Year 12 from 1990*. Retrieved from www.curriculumsupport.education.nsw.gov.au/env_ed/assets/pdf/review_skamp.pdf.

9 Emphasising sustainability when learning power system markets in higher education

William Infante and Jin Ma

Sustainability in higher education

Sustainability is an integral part of decision making that cannot just be sidelined in today's community, and for a holistic consideration, sustainability could be seen in three dimensions: environmental, social, and economic (Kaivo-oja, Panula-Ontto, Vehmas, & Luukkanen, 2014; Krajnc & Glavič, 2005; Riahi, McCollum, & Wiberg, 2014). To bring the focus in disseminating the importance of sustainability, its dimensions and the link between the dimensions, education plays an active and leading role (Hansmann, Mieg, & Frischknecht, 2012). Education for Sustainability (EfS) is at all levels of learning including those in higher education. Although there are efforts to push this role of sustainability to the higher education curriculum, challenges for sustainability are still ongoing whether it be from administrative staff to lecturers (Adomssent, 2006; Davim, Leal Filho, & SpringerLink, 2016). There is also a move for in-depth studies with support from governments promoting sustainability for higher education (Thomas & Millar, 2016).

Sustainability in electrical engineering

Even given the support from different sectors, not all fields in higher education have the same scope or have reached the same rate in terms of integrating concepts of sustainability. The push for EfS has also been given attention in the business school setting. Core concepts gained from EfS in this setting developed individuals to make decisions such that the students would be geared to social action and/or change (Figueiró, Bittencourt, & Schutel, 2016). A similar article has been written about EfS in architecture and built engineering. The author acknowledges the role of higher education to prepare students as promoters for sustainable development (Conte, 2016). This process though should also reach other learners such as those from electrical engineering.

Furthermore, the promotion of EfS in electrical engineering should not be just limited to introductory courses but also to specialisation subjects.

Although there is already an increasing pressure to equip students with real-world skills, specialised courses should not use the importance of advanced or abstract methodologies over the sustainability imperative as an excuse not to incorporate sustainability into their programme. In the long run, it would be beneficial if students can connect specialised subjects learning outcomes to future use of resources. For electrical engineering, creating a way of teaching sustainability becomes crucial as the electricity industry is experiencing a new wave of changes such as the increased use of renewable energies, the deregulation of the electricity industry, and rethinking of businesses associated with the industry (Craig, 2016; Razeghi, Shaffer, & Samuelsen, 2017; Swisher & McAlpin, 2006). Given this scenario, EfS could be realised when teaching power system planning and markets. The aim is to provide a participatory way for students in engineering to rethink models or solutions where the approach has familiarity with the way problems are conventionally taught in engineering. The process needs to involve a process to introduce students to the different aspects of sustainability.

Process-based approach and three EfS pillars

For a holistic approach in teaching about sustainability to engineering students, three pillars have been provided as guides in evaluating a sustainable project; namely the social, environmental and economic aspects (Riahi et al., 2014). In this practice, alongside the holistic approach, our work suggests a process-based approach that could be applied when practising Education for Sustainability in electrical engineering specialised fields. The proposal involves the use of changing landscapes, consideration of stakeholders and a test for robustness.

Each process is then broken down into the three pillars of sustainability to explore the improvement of initial ideas. The application of the framework is presented through operational strategy examples for battery swapping or charging systems in power system markets. Furthermore, the framework can be implemented first in cross-boundary specialised subjects, and later to other core fields in electrical engineering.

Figure 9.1 presents the process-based technique with sustainability dimensions. The process-based techniques are shown in column headers, and spanning the columns are the three aspects of sustainability. The diagram aims to help in the initial assessment if the proposed idea in electrical engineering has the qualities of a sustainable project while also meeting process standards found in engineering. Initially, learners will identify a set of problems in the changing landscapes category, consider the relevance of stakeholders, and finally, learners test the built model for robustness. After implementing an initial process-based technique, the idea is re-evaluated if the proposal can touch more on the three pillars of sustainability, or if the test did not meet the criteria of a certain scenario.

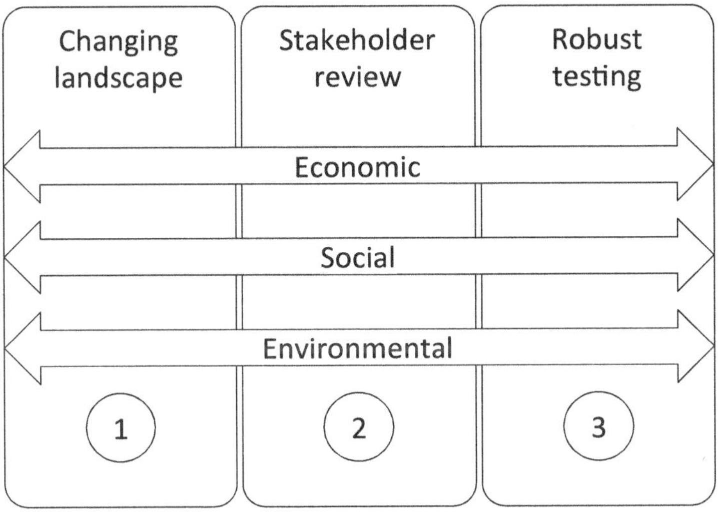

Figure 9.1 Framework for employing sustainability in electrical engineering.

An operational strategy example in the power system markets is provided in the next section.

Learners generate ideas as they start to fill in boxes considering the sustainability dimensions in the diagram. The framework acts as a guide where an idea may not rigidly belong to a certain box in the element. The aim is to provide a dimensional approach to the problem. For example, stakeholder review may have overlapping ideas when evaluating the social impact, but in filling out the diagram, learners may consider a complementary solution or another dimension to the problem. This is not a failure of the approach as the aim is to develop more comprehensive resolutions for the problem. In the given example, the goal is still achieved.

Application in power system markets

To provide an understanding of how to use the process-based approach in conjunction with the three pillars of sustainability, an operational strategy example is provided. As a guide, Table 9.1 is presented showing a summary of the operational strategy example for battery swapping or charging systems in the power system markets. In Table 9.1, aspects of sustainability, as well as a process-based approach practised in engineering, are combined to provide a comprehensive way of formulating or reformulating a problem and providing possible solutions. It is likely that the initial process will only fill-out a few of the boxes in the table, but as the idea develops, more boxes are filled. For example, initially, changing landscapes would only include electricity deregulation and the integration of the renewable energies, but

Table 9.1 Operational strategy examples in power system markets

Aspects/process	*Changing landscapes*	*Stakeholder consideration*	*Testing for robustness*
Economic	• Electricity deregulation • Subsidies provided to renewable energy companies	• EV customers: observing the role of incentives • BSS owner: how supplementary services can be beneficial, using optimisation techniques for computing net value • Grid operators: how grid ancillary services can help support intermittent renewable energy integration	• Potential profits during low and high EV market penetration • Operational strategy for scenarios with and without government subsidies • Comparison of the overall cost of owning an electric vehicle compared to a conventional vehicle (including gasoline and electricity price changes)
Social	• Increased importance of demand response and its effect on consumers • New provisions in a power system market opening roles for non-traditional businesses	• Multi-objective optimisation of maximising the benefits of BSS owners and electric vehicle customers • Consideration of the response of electric vehicle customers to incentives	• Test scenarios when the relative percentage of cooperative and indifferent customer changes
Environmental	• Harmful impact of the early disposal of Lithium-ion batteries • Increased integration of renewable energy sources	• Second-life applications of electric vehicle batteries • During the disposal phase, disposal facility operator plays a key role	• Scenario testing with optimally available batteries for EV use • Scenario re-adjustments for inclusion of battery state of health in battery allocation

after trying to fill-out the table, ideas added to the changing landscape included subsidies given by the government as well as the impact of lithium-ion battery disposal creating a more comprehensive approach to the problem. The next sections would discuss the activities per process.

Ideas are presented on the changing landscape in the power system market. Following this, a review of the new stakeholders is emphasised. Scenarios are then tested in the final process. For the last part, robust testing is usually repeated multiple times as adjustments are made from partial results of the test to improve the current model.

Changing landscape

This section aims to provide an overview of the changing landscape and how these are used to trigger opportunities for learners to rethink possible implementations.

Changing landscape: economic aspect

The power system has been undergoing electricity deregulation, in different places around the world and this process continuously explores market mechanisms (Biggar & Hesamzadeh, 2014). There are different motives for electricity deregulation from different stakeholders. For example, for the end user and the government, it aims to reduce the prices and foster customer choice. For those in the transmission and operators, it aims to increase efficiency and foster innovation through competition.

Ideally, the electricity demand should always meet the supply and be at a manageable level. If not, load shedding can occur where it could create an interruption in the electricity supply to customers. Another approach is to set up additional facilities to accommodate unexpected demands translating to higher electricity costs.

To provide learners with an understanding of the change in electricity distribution, the demand for electricity changes throughout the year is presented in Figure 9.2. It represents the electricity demand for New South Wales for 2016 with high demand during winter and summer seasons.

Furthermore, in the recent years, government subsidy has been provided for projects related to the feasibility and deployment of renewable energy sources (Yang & Nie, 2016; Yu, Guo, Le-Nguyen, Barnes, & Zhang, 2016; Zhang, Li, Zhou, & Zhou, 2014). An increase then of potential renewable sources should be considered when looking for a potential solution.

Changing landscape: social and environmental aspects

Traditionally, market entities in the electricity market are very limited. Sometimes, there is only one independent system operator. Now, with

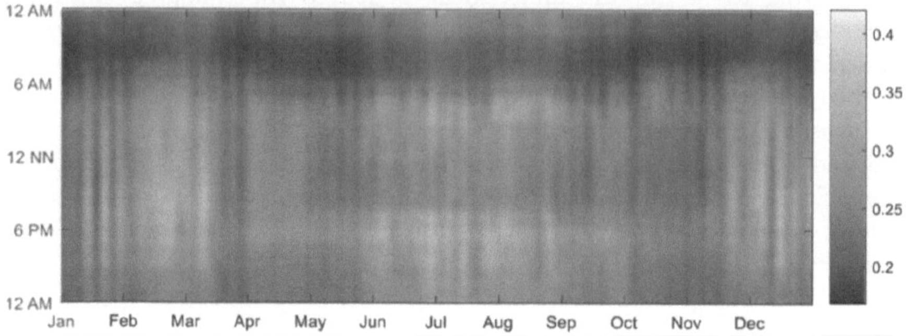

Figure 9.2 Relative electricity demand in New South Wales, Australia for 2016 showing peak electricity demand during summer and winter seasons.

deregulation, there are more possible market participants. The independent system operator can be separated from the generating company and the transmission company. Moving forward, there could even be more transmission companies. New electricity retail companies can be formed where this new entity in the electricity market gains legal approval to sell and retail electricity. It now buys electricity or other services to provide electricity and present them in different packages. Participants are not just limited as entities, as aggregators can also participate (Rothwell & Gomez, 2003; Stoft, 2002). There are now more companies that participate in the electricity market. This also provides competition where businesses can be created with a more sensitive demand response; customers are empowered with the choices that are provided to them.

The push for the integration of renewable energies also opened opportunities for new businesses. However, this has also caused a changing landscape that brought problems in the electricity load demand because of the intermittency of renewable energies (Brouwer, Van Den Broek, Seebregts, & Faaij, 2014). If we try to visually track the highest and lowest electricity demand per day for the whole year in two different regions in Australia as in Figure 9.3, noticeably that there are more off-peak values during midday in South Australia compared to New South Wales. Based on the Australian Energy Market Operator report, one of the causes in the shift of low-demand values was the large penetration of rooftop solar panels in homes (AEMO, 2015). Homes demanded less electricity at midday as they could use electricity from the solar panels. Visually from the graph, electricity demand in the South Australia pattern for the peaks is not as regular as that of New South Wales. From the market electricity pricing, there are even instances when there is negative electricity demand in South Australia due to the over generation of electricity. This means that consumers are paid to consume electricity (Chaiamarit & Nuchprayoon, 2014).

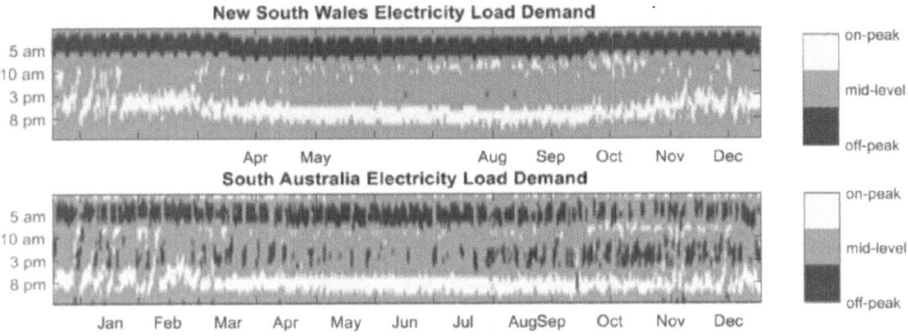

Figure 9.3 Difference in on-peak and off-peak electricity load demand of two regions in Australia with different rooftop solar panels.

In another region, a similar issue is observed in terms of renewable energy generation, the abundance of solar photovoltaic in California produced too much energy at noon time causing higher generation in the electricity infrastructure (Razeghi et al., 2017). In electricity infrastructure, it is desirable to have an almost constant electricity load so that the infrastructures can be operated optimally in the prescribed range or conditions. Having variable power means more backups to be installed to make up for the low power generation and additional storage facilities for high peaks.

There may be long-run economic benefits in producing electricity from renewable energies and environmental benefits as it is a potential alternative to fossil fuels, but without an efficient operational strategy, this could also lead to higher electricity infrastructure costs passed on to the consumers. If not managed, this creates undesirable social costs.

During the initial phase of looking for a solution, the proposed solution of using electric vehicle battery swapping or charging stations provided another sustainable dimension in the changing landscape. Disposal of batteries was identified as an environmental issue in the deployment. The model was then revised to complement the electric vehicle battery swapping station with ancillary services so that the batteries can have second use even after reaching their service life from electric vehicle battery swapping. For example, the batteries could still be used for grid services before disposing of them (Neubauer & Pesaran, 2011; Viswanathan & Kintner-Meyer, 2011). The modified model was meant to create an opportunity to increase profits as well as create a sustainable environment by maximising the resources available.

The environmental aspect may also be linked to the economic aspect in terms of the disposal costs. Disposal may be considered a cost in the first instance, but as battery disposal technology improves, the disposal may even lead to a positive net income for the business owner, and it also benefits the environment (Wang, Gaustad, Babbitt, & Richa, 2014).

All these changes provide a platform to tackle the problem and provide sustainable solutions. Honing of students' understanding through questions in the field of engineering is particularly important. The changing landscapes make learners think not only in terms of just one business, but in terms of situating the context for thinking of models in terms of the different aspects of sustainability.

Relevant stakeholder review

Relevant stakeholder review: economic aspect

During this phase, learners identify the beneficiaries and prime movers with the changing landscape. In the provided operational strategy, the battery swapping station owner becomes a non-traditional stakeholder in the system. Battery swapping station owners aim to maximise their profit while using incentives and allocating batteries based on electricity pricing. The customers, although former players in the traditional industry, have now more prominence as their behaviour in using services influences the business outcome. Their aggregated behaviour through the battery swapping and charging station with grid services can make them a comparable stakeholder in the system. Grid operators are also stakeholders as they benefit from the participation in ancillary services by battery swapping stations.

Relevant stakeholder review: social aspect

The stakeholder review overlaps with the social perspective, but it is important to note here that the aim of the different stakeholders may be contradictory. For example, battery swapping station owners want to maximise their profit, while customers want to increase their savings. Maximising an objective of one stakeholder may not necessarily maximise another stakeholder. For this case, in creating an operational strategy for battery swapping stations, a multi-objective strategy should be considered for the model.

Based on this example, the model has been slowly moving from a conventional battery swapping station, into a station with complementary services, and into a station with complementary services that has a multi-objective operational strategy.

Relevant stakeholder review: environmental aspect

At different phases of the battery swapping station business, different stakeholders may be present. For example, as we have already considered the battery disposal in the environmental phase during the changing environment, additional stakeholders from the battery disposal facility may be present. The battery swapping station not only deals with the electricity sector, but it also has an interface with the electric vehicle sector. Although these may not be the primary stakeholders, their interaction may be significant at different stages.

Test for robustness

Test for robustness: economic aspect

The operational model has been generated based on the changing landscape, and its interaction with stakeholders. The process is continued by testing it with scenarios to ensure a sustainable outcome. If the model, for example, does not meet the criteria, a review of the model is made. If the initial solution is a conventional battery swapping station, it may not be feasible when there are a low number of electric vehicle penetrations. The rate of electric vehicle users may also decrease as gasoline and conventional vehicle prices are reduced. Having a supplementary service such as a grid service can be an additional stream of income even with low electric vehicle penetration. In addition, using such a service could also be helpful for the environment as the batteries can still be used even before disposal.

Test for robustness: social and environmental aspects

The participation of electric vehicle customers due to incentives provided by the battery swapping or charging station may also influence the sustainability of the business. The scenarios of different cooperative and indifferent customers may be helpful in creating a robust model.

In testing the possibility of using grid services for battery services, scenarios involving battery state of health would help create new ways to extend battery service. Modifying the electricity load demand could also check the sensitivity of the model to different loads such as those with high rooftop solar panel penetration, and those with significant amounts of renewable energy sources.

Conclusion

This chapter discussed how the concept of a process-based approach with the pillars of sustainability is used to introduce Education for Sustainability through concepts in electrical engineering. Learners, given the framework, can reconsider the current landscape and where it is heading, and situate the landscape in an ecosystem for opportunities or model solutions that benefit the stakeholders. The work needs constant revision as a test for robustness would mean modifying or constraining the operational strategy to meet the objective of the business. In addition, after initially creating a process-based approach, the solution is then re-evaluated in terms of the three pillars of sustainability. This again adds rounds of revision, but the solution generated becomes more comprehensive. In this way, the framework creates a participatory atmosphere with a process familiar to those in the engineering practice of finding solutions to real-world problems and testing them for robustness. The method also looks at the economic, social and environmental aspects. The learners are not just introduced to new concepts of thinking but leverage on the practice they are familiar with. This leads to a thinking or mindset of implementing sustainability in the learners even in the decisions they make in their future careers.

Recommendations and future work

This chapter shows that by starting with what is already familiar to engineering students, we can move to introduce sustainability concepts. The power systems market course could be a prime candidate for the introduction of Education for Sustainability as the electricity industry is undergoing rapid change. This is classified as a cross-boundary subject needing a more exploratory way of thinking. Cross-boundary subjects use techniques from one field and borrow from other fields. For power system markets, its main core is understanding the electricity industry from the electrical engineering background but also borrows methods from behavioural and energy economics. Eventually, this process could be translated to other specialised subjects. The framework that was introduced has elements of a process-based approach and aspects of sustainability. Realignment of topics by providing learners with some familiarity for tackling problems and then expanding this with the three aspects of sustainability could be a practical tool for the introduction of EfS in electrical engineering.

References

Adomssent, M. (2006). *Higher Education for Sustainability: New Challenges from a Global Perspective (Vol. 1)*. Frankfurt am Main: VAS.

Australian Energy Market Operator (AEMO). (2015). *South Australian Electricity Report*. Sydney: AEMO.

Biggar, D. R., & Hesamzadeh, M. R. (2014). *The Economics of Electricity Markets*. Hoboken: Wiley.

Brouwer, A. S., Van Den Broek, M., Seebregts, A., & Faaij, A. (2014). Impacts of large-scale intermittent renewable energy sources on electricity systems, and how these can be modeled. *Renewable and Sustainable Energy Reviews, 33*, 443–466. doi:10.1016/j.rser.2014.01.076.

Chaiamarit, K., & Nuchprayoon, S. (2014). Impact assessment of renewable generation on electricity demand characteristics. *Renewable and Sustainable Energy Reviews, 39*, 995. doi:10.1016/j.rser.2014.07.102.

Conte, E. (2016). Sustainability and built environment: The role of higher education in architecture and building engineering. *European Journal of Sustainable* Development, 5(3), 1. doi:10.14207/ejsd.2016.v5n3p1.

Craig, J. D. (2016). Motivations for market restructuring: evidence from US electricity deregulation. *Energy Economics, 60*, 162–167.

Davim, J. P., Leal Filho, W., & SpringerLink. (2016). *Challenges in Higher Education for Sustainability* (Vol. 1). Cham: Springer International Publishing.

Figueiró, P. S., Bittencourt, B. A., & Schutel, S. (2016). Education for sustainability in business schools by practicing social learning. *Brazilian Journal of Science and Technology, 3*(1), 11. doi:10.1186/s40552-016-0014-7.

Hansmann, R., Mieg, H. A., & Frischknecht, P. (2012). Principal sustainability components: Empirical analysis of synergies between the three pillars of sustainability. *International Journal of Sustainable Development & World Ecology, 19*(5), 451–459. doi:10.1080/13504509.2012.696220.

Kaivo-oja, J., Panula-Ontto, J., Vehmas, J., & Luukkanen, J. (2014). Relationships of the dimensions of sustainability as measured by the sustainable society index framework. *International Journal of Sustainable Development & World Ecology, 21*(1), 39–45. doi:10.1080/13504509.2013.860056.

Krajnc, D., & Glavič, P. (2005). How to compare companies on relevant dimensions of sustainability. *Ecological Economics, 55*(4), 551–563.

Neubauer, J., & Pesaran, A. (2011). The ability of battery second use strategies to impact plug-in electric vehicle prices and serve utility energy storage applications. *Journal of Power Sources, 196*(23), 10351–10358. doi:10.1016/j.jpowsour.2011.06.053.

Razeghi, G., Shaffer, B., & Samuelsen, S. (2017). Impact of electricity deregulation in the state of California. *Energy Policy, 103*, 105–115. doi:10.1016/j.enpol.2017.01.012.

Riahi, K., McCollum, D., & Wiberg, D. (2014). Prototype global sustainable development report. Policy Analysis Branch of the Division for Sustainable Development, UN Department for Economic and Social Affairs (DESA), New York, USA. Retrieved from http://pure.iiasa.ac.at/id/eprint/11145/.

Rothwell, G., & Gomez, T. (2003). Electricity economics: Regulation and deregulation. *The Energy Journal, 24*(3), n.p.

Stoft, S. (2002). *Power System Economics: Designing Markets for Electricity*. Piscataway, NJ and New York: IEEE Press.

Swisher, J. N., & McAlpin, M. C. (2006). Environmental impact of electricity deregulation. *Energy, 31*(6), 1067–1083. doi:10.1016/j.energy.2005.02.006.

Thomas, I., & Millar, S. (2016). Sustainability, education and local government: Insights from the Australian state of Victoria. *Local Environment, 21*(12), 1482–1499. doi:10.1080/13549839.2016.1140131.

Viswanathan, V. V., & Kintner-Meyer, M. (2011). Second use of transportation batteries: Maximizing the value of batteries for transportation and grid services. *IEEE Transactions on Vehicular Technology, 60*(7), 2963–2970. doi:10.1109/TVT.2011.2160378.

Wang, X., Gaustad, G., Babbitt, C. W., & Richa, K. (2014). Economies of scale for future lithium-ion battery recycling infrastructure. *Resources, Conservation and Recycling, 83*, 53–62. doi:10.1016/j.resconrec.2013.11.009.

Yang, D.-x., & Nie, P.-y. (2016). Influence of optimal government subsidies for renewable energy enterprises. IET Renewable Power Generation. doi:10.1049/iet-rpg.2015.0307.

Yu, F., Guo, Y., Le-Nguyen, K., Barnes, S. J., & Zhang, W. (2016). 'The impact of government subsidies and enterprises' R&D investment: A panel data study from renewable energy in China. *Energy Policy, 89*, 106–113. doi:10.1016/j.enpol.2015.11.009.

Zhang, H., Li, L., Zhou, D., & Zhou, P. (2014). Political connections, government subsidies and firm financial performance: Evidence from renewable energy manufacturing in China. *Renewable Energy, 63*, 330–336. doi:10.1016/j.renene.2013.09.029.

10 (How) do students reflect on sustainability?

A model to diagnose and foster reflective thinking about sustainability

Nina Brendel

Introduction

Building awareness for sustainability is a key challenge of modern (geography) education: international guidelines such as the UNESCO Roadmap (UNESCO, 2014) or the UNESCO Learning Objectives (Rieckmann, 2017) as well as national frameworks and reports (e.g. KMK & BMZ, 2016, Smith, Rowe, & Vorva, 2015; UK National Commission for UNESCO, 2013; UNESCO, 2006) encourage educators worldwide to implement issues of sustainability into their teaching and foster competencies that enable students to face the challenges of global change.

In order to accomplish these goals, Education for Sustainable Development (ESD) – a concept closely related to Education for Sustainability (EfS) (Delgado, 2007) – highlights the significance of reflective thinking as a key competence for learners and educators in ESD (Rieckmann, 2017; UNECE Strategy for ESD, 2012). In contrast to the high importance that is attached to processes of student reflection in ESD, little is known about how reflective thinking can be identified, influenced or increased in the classroom. A 'reflexive engagement with ESD' (Buckler & Creech, 2014, p. 32) is still seen as a challenge. No empirical studies were located relating to how these reflective processes of secondary education students can be determined and how educators can facilitate reflective thinking on sustainability and thus foster capacity-building to address concerns of sustainable development with their students (United Nations, 1992).

Therefore, the objective of this study is to address this need by developing an empirical multi-stage model designed to help educators diagnose different levels of reflective thinking and to increase the reflective performance of their students. Furthermore, the chapter presents factors that influence reflective performance in the context of sustainability. In doing so, the chapter presented here specifically focuses on findings of this study relevant for ESD and EfS. For findings on the power of geographical thinking see Brendel (2017a & 2017b).

Theoretical background

The implementation of ESD in educational settings rests on three key pedagogical approaches (Rieckmann, 2017, p. 55): a learner-centred approach, action-oriented learning and transformative learning. Each of these approaches explicitly states processes of reflection to be a core element of the concept: first, a learner-centred approach 'require[s] learners to reflect on their own knowledge and learning processes in order to manage and monitor them. Educators should stimulate and support those reflections' (Rieckmann, 2017, p. 55). Second, action-oriented learning draws on the experiential cycle by Kolb, which argues that construction and application of concepts are based on the reflection of experience (ibid.). Third, transformative learning aims at empowering learners to question and change the ways they see and think about the world in order to deepen their understandings (Slavich & Zimbardo, 2012; Mezirow, 2000; Rieckmann, 2017).

From this, we can draw two conclusions: on one hand, reflective thinking seems to be a key issue of pedagogies for ESD (ibid.). On the other hand, the three approaches mentioned above refer to different concepts of reflection, e.g. proposed by Kolb (1984), Mezirow (2000) or Dewey (1933). While these theories show a lot of parallels, built upon or influenced by each other, they still differ in focus (Atkins &Murphy, 1993). So although the relevance of reflection processes in ESD is indisputable, it is worth examining more closely what is meant by reflection in the context of ESD.

In this study, two types of reflection processes are distinguished (Hatton & Smith, 1995): those that draw on the concept of 'reflective thinking' as defined by Dewey (1933, 1997) and those that build on the concept of 'reflective practice' as defined by Schön (1983). By definition, reflective thinking is an 'active, persistent, and careful consideration of any belief or supposed form of knowledge in the light of the grounds that support it, and the further conclusions to which it tends' (Dewey, 1997, p. 6). Reflective practice, on the other hand, implies two sides of reflection: reflection-in-action (intuitive and sometimes unconscious knowledge of how to do something *while* doing it) and reflection-on-action (reflecting on actions in the past and their consequences) (Schön, 1983; Sellars, 2014).

In a nutshell, reflective thinking refers to deep thinking about *content*, while reflective practice is concerned with thinking about *actions, their alternatives and consequences*. Both concepts have been used extensively in the literature in order to determine reflection processes of students and educators (Bain, Ballantyne, & Packer, 1999; Chen, Wei, Wu, & Uden, 2009; Henderson, Napan, & Monteiro, 2004; Hsieh, Jang, Hwang, & Chen, 2011; King & Kitchener, 2004).

In this study, the aim is to understand and analyse the extent to which students of secondary education reflect on issues of sustainability such as sustainable water management, tourism or consumption. Therefore, this study considers the reflection of the content and is thus based on Dewey's concept of reflective thinking. The prime research questions are: which levels of reflective thinking do students reach in the context of global learning and which factors influence these reflection processes?

Method

In order to determine the levels of reflective thinking attained by students of secondary education, four geography classes (grades 10 and 11, students aged 15–17, German high schools) studying sustainability over the course of 4 to 8 weeks were followed.

The aim was to analyse the students' reflective thinking performance in an everyday learning setting, therefore the study refrained from developing and applying a best-practice unit. The study opted for analysing authentic geography instruction as it takes place in schools every day. For this reason, this study is based on units on sustainability which were fully in the responsibility of the teachers and in accordance with the curriculum; the author did not attend the classes but stayed in the background as much as possible. Still, creating a learning environment which facilitates reflective thinking was crucial for the project. Some studies have proven weblogs to be an adequate method of fostering reflective writing (Lin & Yuan, 2006; Xie, Ke, & Sharma, 2008; Yang, 2009), which is why all 64 students were to reflect on each lesson's contents in a personal weblog and comment on their peers' weblog articles. Furthermore, the teachers were asked to reflect on their lessons, their instruction and on the class' responses in a personal weblog, which was only accessible to themselves and the author. In addition, semi-structured qualitative interviews with the teachers prior to the project gained an insight into the learning environment and classroom situation as well as the teachers' perception of reflection, ESD and use of weblogs as a learning tool. Figure 10.1 depicts the research design.

In order to analyse the students' levels of reflective thinking, the work of Dewey (1933) and pre-existing multi-stage models of reflective thinking (Bain et al., 1999; Chen et al., 2009) were used as a guide. Using a qualitative, iterative data analysis approach based on variations of content analysis by Mayring (2000), a multi-stage model to be applicable in ESD was created (see Figure 10.2).

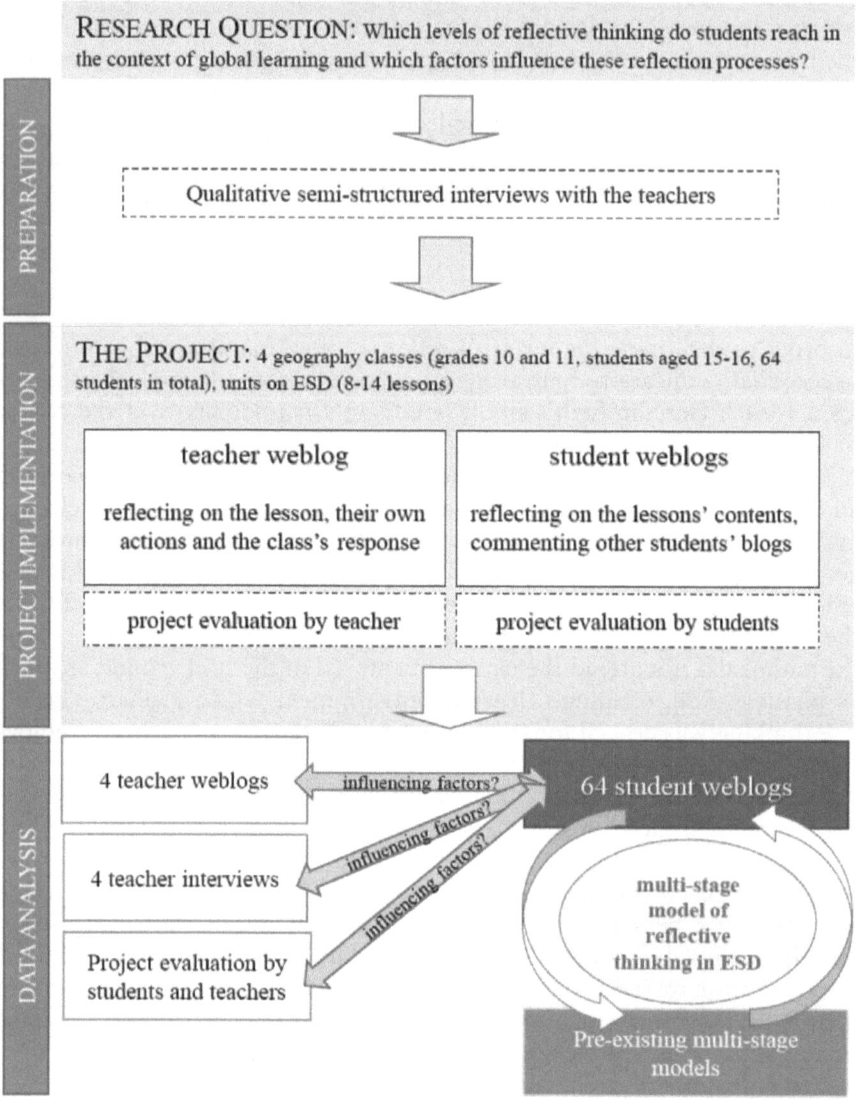

Figure 10.1 The research design.

Level of reflective thinking		Criteria to diagnose each level
Level 1: reporting		Pure reproduction of contents (no reflection)
Level 2: responding		Learners: -slightly transform or conceptualise the lesson's content -Utter an opinion or observation, but without giving a reason or conclusion -Ask rhetorical questions without giving answers -Express a feeling -State something they did not know before
Level 3: relating	3a: giving reasons	Learners: -Give superficial reasons for processes or issues, realise a need to act -Express a judgement and provide a simple reason
	3b: connecting	Learners: -Somehow connect contents to personal experience or prior knowledge -Try to outline superficial interrelations
Level 4: assessment		***Prerequisite: conceptualisation on higher level*** Learners: -Link content to theory or personal experience -Give more sophisticated reasons for processes or issues or actions -Analyse a problem, search for answers and alternatives, speculate or state a hypothesis -Investigate the relation of theory and practice -Realise the complexity of an issue
Level 5: (re)construction		***Prerequisite: conceptualisation on high level*** Learners: -Draw personal conclusions, take a sophisticated stand and think in systems -Formulate a personal theory or derive a generalisation -Extend and modify the results of the lesson significantly beyond the target level

Figure 10.2 Multi-stage model of reflective thinking of secondary education students.

Source: Brendel, 2017a, translated.

Selected findings

Level of students' reflective thinking on sustainability

In the course of the analysis of the student weblogs, teacher weblogs and project evaluation using this multi-stage model of reflective thinking, differences in the students' reflection levels were found: some students' reflections were limited to merely reporting or describing the lessons' contents, others, however, reflected at higher levels by giving detailed reasons and conceptualising what they had learned. Figure 10.3 illustrates this by comparing the levels of reflective thinking achieved by three different students of class A.

Considerable differences were also found among lessons of the same unit: some lessons triggered only low levels of reflection (e.g. lesson 7, Figure 10.4) while other lessons leveraged high levels of student reflection (e.g. lessons 5 and 11, Figure 10.4).

Factors influencing reflection on sustainability

In every unit of the four classes, considerable differences were observed in the students' reflective thinking performance, among students as well as among lessons. However, it was even more interesting and relevant for this study to identify the reasons and underlying process responsible for these effects, as the factors that influence reflective thinking about sustainability were sought. Overall, the data analysis detected eight influencing factors:

- the extent to which content is connected to the students' everyday life;
- how sustainability is dealt with in geographic content;
- the way of communicating;

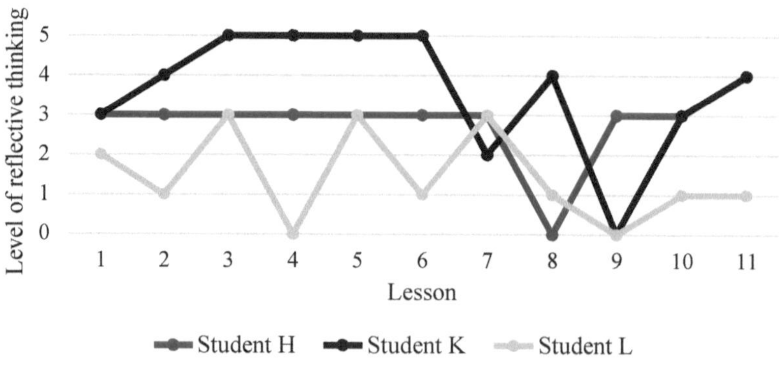

Figure 10.3 Comparison of levels of reflective thinking achieved by three different students of class A.

Source: based on Brendel, 2017a.

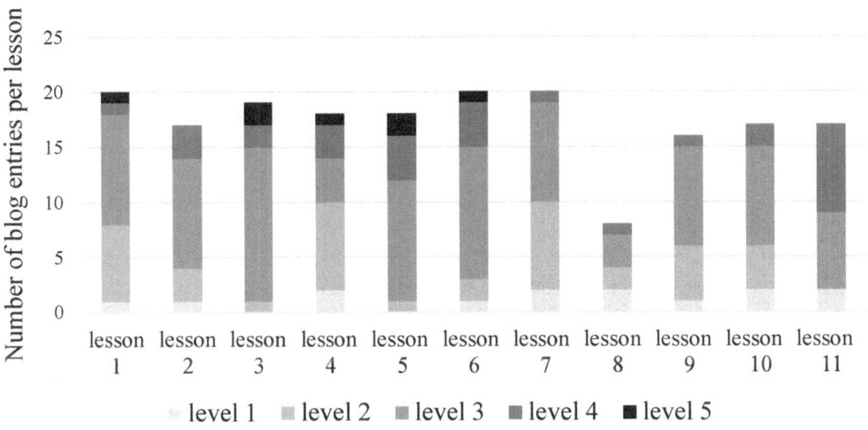

Figure 10.4 Levels of reflective thinking of all students of class A.

- the level of student participation;
- pedagogies of teaching sustainability;
- the actions of the teacher;
- the learning environment, and
- gender.

In this chapter we highlight three results, which have particular impacts for ESD/EfS implementation and instruction in secondary schools (Brendel, 2017a):

Students' reflective thinking performance improves if…

1 The lesson content is linked to the students' everyday lives. This is by far the most important factor on reflective thinking performance and is particularly relevant to female students and teachers. Bearing in mind that many issues of ESD have little personal meaning for students in Central Europe – e.g. the sustainable development goals 'no hunger' or 'clean water and sanitation' (United Nations, 2016) – it is crucial for teachers to establish links to the students' lifeworld to increase their reflective thinking.
2 Dimensions of sustainability (ecology, economy, society and politics) and levels of scale are not studied in isolation, but discussed with a focus on the interrelations and mutual influence. Often this entails a higher degree of complexity, which can be a challenge for lower achieving students. However, the findings indicate a connection between competencies to deal with complexity and higher levels of reflective thinking.
3 Teachers challenge students to actively solve problems by themselves. While this might result in a feeling of uncertainty and lower reflective

thinking performance at first, reflective levels increase, when students overcome their initial insecurities and begin to solve problems autonomously.

Furthermore, the data analysis revealed several triggers (reflection prompts, certain teaching methods, etc.) that can help teachers to individually promote students' reflection levels based on their dominating level of reflective thinking performance (Brendel, 2017a).

Conclusion: ways to foster student reflection on sustainability

In light of these results, there are several conclusions for facilitating reflective thinking about sustainability in secondary education:

1. The findings highlight the importance of emphasising the interrelations and interdependencies of multiple levels of scale and all dimensions of sustainability.
2. There is an argument for establishing links to the students' everyday lives whenever possible, in order to increase reflective thinking performance.
3. The results suggest that confronting students with the complexity inherent in issues of sustainability and challenging them to deal with uncertainty is a necessity.
4. Based on the analysis of the data (student weblogs, teacher weblogs, teacher interviews and evaluations), we assume a correlation between elevated levels of reflective thinking and a deeper understanding of sustainability still requires further investigation.

Due to the qualitative approach of this study, it is important to stress its limitations. Further studies have yet to investigate if the results are valid for other learning settings and if the findings are applicable to other case studies. Nevertheless, the outcomes of this study are valuable for educators in high school and higher education, who seek to diagnose students' reflective thinking. Educators can use the multi-stage model and adapt their instruction with respect to the influencing factors of reflective thinking about sustainability.

References

Atkins, S., & Murphy, K. (1993). Reflection: A review of the literature. *Journal of Advanced Nursing*, *18*, 1188–1192.

Bain, J. D., Ballantyne, R., & Packer, J. (1999). Teachers and teaching: Theory and practice using journal writing to enhance student teachers' reflectivity during field experience placements. *Teachers and Teaching: Theory and Practice, 5*(1), 51–73.

Brendel, N. (2017a). *Reflexives Denken im Geographieunterricht: Eine empirische Studie zur Bestimmung von Schülerreflexion mithilfe von Weblogs im Kontext Globalen Lernens*. Dissertation. *Erziehungswissenschaft und Weltgesellschaft: Band 10*. Münster: Waxmann.

Brendel, N. (2017b). Using Weblogs to Determine the Levels of Student Reflection in Global Education. In Brooks, C., Butt, G., & Fargher, M. (Eds.), *International Perspectives on Geographical Education. The Power of Geographical Thinking* (pp. 119–135). Cham: Springer.

Buckler, C., & Creech, H. (2014). *Shaping the Future We Want: UN Decade of Education for Sustainable Development (2005–2014): Final Report. DESD Monitoring and Evaluation*. Paris: UNESCO.

Chen, N.-S., Wei, C.-W., Wu, K.-T., & Uden, L. (2009). Effects of high-level prompts and peer assessment on online learners' reflection levels. *Computers and Education, 52*(2), 283–291. https://doi.org/10.1016/j.compedu.2008.08.007.

Delgado, L. (2007). *Education for Sustainability in Local Government: Handbook*. Retrieved from http://aries.mq.edu.au/handbook/.

Dewey, J. (1933). *How We Think: A Restatement of the Relation of Reflective Thinking to the Educative Process*. Boston: D. C. Heath and Company.

Dewey, J. (1997). *How We Think*. Mineola: Dover Publications.

Hatton, N., & Smith, D. (1995). Reflection in teacher education: Towards definition and implementation. *Teaching and Teacher Education, 11*(1), 33–49. Retrieved from www.sciencedirect.com/science/article/pii/0742051X9400012U.

Henderson, K., Napan, K., & Monteiro, S. (2004). Encouraging Reflective Learning: An Online Challenge. In Atkinson, R., McBeath, C., Jonas-Dwyer, D., & Phillips, R. (Eds.), *Beyond the Comfort Zone: Proceedings of the 21st ASCILITE Conference* (pp. 357–364). Perth: ASCILITE.

Hsieh, S.-W., Jang, Y.-R., Hwang, G.-J., & Chen, N.-S. (2011). Effects of teaching and learning styles on students' reflection levels for ubiquitous learning. *Computers and Education, 57*(1), 1194–1201. https://doi.org/10.1016/j.compedu.2011.01.004.

King, P. M., & Kitchener, K. S. (2004). Reflective judgment: Theory and research on the development of epistemic assumptions through adulthood. *Educational Psychologist, 39*(1), 5–18. https://doi.org/10.1207/s15326985ep3901_2.

KMK, & BMZ (Eds.). (2016). *Orientierungsrahmen für den Lernbereich Globale Entwicklung im Rahmen einer Bildung für nachhaltige Entwicklung: Ergebnis des gemeinsamen Projekts der Kultusministerkonferenz (KMK) und des Bundesministeriums für wirtschaftliche Zusammenarbeit und Entwicklung (BMZ)* (2te aktualisierte und erweiterte Auflage). Berlin: Cornelsen Verlag.

Kolb, D. A. (1984). *Experiential Learning: Experience as the Source of Learning and Development*. Englewood Cliffs, NJ: Prentice-Hall.

Lin, H. T., & Yuan, S. M. (2006, July). Taking Blog as a Platform of Learning Reflective Journal. In *International Conference on Web-Based Learning* (pp. 38–47). Berlin: Springer.

Mayring, P. (2000). Qualitative content analysis. *Forum Qualitative Sozialforschung/Forum: Qualitative Social Research, 1*(2), n.p.

Mezirow, J. (2000). *Learning as Transformation: Critical Perspectives on a Theory in Progress* (1st ed.; The Jossey-Bass Higher and Adult Education Series). San Francisco, CA: Jossey-Bass.

Rieckmann, M. (2017). *Education for Sustainable Development Goals: Learning Objectives*. Paris: UNESCO. Retrieved from http://unesdoc.unesco.org/images/0024/002474/247444e.pdf.

Schön, D. A. (1983). *The Reflective Practitioner: How Professionals Think in Action*. New York: Basic Books. Retrieved from www.loc.gov/catdir/enhancements/fy0832/82070855-d.html.

Sellars, M. (2014). *Reflective Practice for Teachers*. Los Angeles: SAGE.

Slavich, G. M., & Zimbardo, P. G. (2012). Transformational teaching: Theoretical underpinnings, basic principles, and core methods. *Educational Psychology Review, 24*(4), 569–608.

Smith, K., Rowe, D., & Vorva, M. (2015). *The Status of Education for Sustainable Development (ESD) in the United States: A 2015 Report to the US Department of State*. Retrieved from www.sustainabilityprofessionals.org/sites/default/files/ESD%20in%2 0the%20United%20States%20final.pdf.

UK National Commission for UNESCO. (2013). *Education for Sustainable Development (ESD) in the UK – Current Status, Best Practice and Opportunities for the Future* (Policy Brief). Retrieved from www.unesco.org.uk/wp-content/uploads/2015/03/Brief-9-ESD-March- 2013.pdf.

United Nations. (1992). *Agenda 21. Rio de Janeiro*. Retrieved from https://sustainabledevelopment.un.org/content/documents/Agenda21.pdf.

United Nations. (2016). Sustainable development knowledge platform. Retrieved from https://sustainabledevelopment.un.org/topics/sustainabledevelopmentgoals.

United Nations Economic Commission for Europe (UNECE Strategy for ESD). (2012). *Learning for the Future: Competences in Education for Sustainable Development*. Retrieved from www.unece.org/fileadmin/DAM/env/esd/ESD_Publications/Competences_Publication.pdf.

United Nations Educational, Scientific and Cultural Organization (UNESCO). (2006). *Pacific Education for Sustainable Development Framework: Endorsed by the Pacific Ministers of Education 27 September 2006, Nadi FIJI*. Retrieved from http://unesdoc.unesco.org/images/0014/001476/147621E.pdf.

United Nations Educational, Scientific and Cultural Organization (UNESCO). (2014). *Roadmap for Implementing the Global Action Programme on Education for Sustainable Development*. Retrieved from http://unesdoc.unesco.org/images/0023/002305/230514e.pdf.

Xie, Y., Ke, F., & Sharma, P. (2008). The effect of peer feedback for blogging on college students' reflective learning processes. *The Internet and Higher Education, 11*(1), 18–25. https://doi.org/10.1016/j.iheduc.2007.11.001.

Yang, S.-H. (2009). Using blogs to enhance critical reflection and community of practice. *Educational Technology & Society, 12*(2), 11–21.

Part III

From theory to practice (translating)

11 Geographies of Education for Sustainability (EfS)

Shaping the EfS in Vietnam's approach to education

Nguyen Minh Quang

Introduction

The last decade has witnessed strong economic growth and an increasingly global integration of economic activity in Vietnam. The country remains determined to move forward with ambitious plans to attract foreign investment, creating new industries and developing its infrastructure to continue its economic development. Having experienced a 'golden population structure' (Dan, 2018) with more than half of its people of working age, it is crucial for Vietnam to develop a well-trained labour force, which needs to be equipped with scientific, technological and management skills.

As new industries expand, a university degree is indispensable for young Vietnamese workers searching for higher paying jobs in newly emerging industries. However, the government has acknowledged that the current education system is unable to meet demand. According to the Vietnamese state media, every year there are as many as 90 per cent of students in Vietnam wanting to enrol in a university (Minh Duong, 2016). Although the number of university students has doubled since 1990, opportunities for higher education, in practice, are limited since the system can accommodate only a fraction of those seeking admission.

More challenging is the fact that Vietnamese universities are not producing the educated workforce that Vietnam's economy and society demand. A survey conducted in 2011 by the Vietnamese Ministry of Education and Training (MOET) found that a large percentage of university graduates face challenges securing jobs in their area of specialisation and many require further on-the-job training courses to ensure their productivity in the workplace (Kim Ngan, 2012). This situation demonstrates that the disconnect between the classroom and the needs of the market is large, thus highlighting the need for a more practical and effective education for students in the country.

Furthermore, the challenges of the Vietnamese education system remain unsettled, highlighted by 'outdated' curricula, teacher-centred 'chalk and talk' methods, and a pedagogical philosophy that stresses abstract concepts rather than practical and soft skills at all levels (Horn, 2014; Ngoc, 2016).

As a result, the decades-long educational reforms have been inadequate in turning Vietnamese youth into a competitive workforce in the region.

Given such a dire situation, MOET mobilised a series of education reforms in terms of governance systems, curriculum, mission focus, external relations, research and financing (Viet, 2009; Thuy Linh, 2015). As a result, the last two decades have seen a drastic change in the education system. However, in a country where the government exercises tight control over social and economic development strategies, wide-scale changes to an education system are often complex and difficult to implement. In response to Vietnam's transition to a socialist-oriented market economy, educational policy planners have had to consider strategies for making the education system more responsive to current and future labour demands. Yet, it still remains unable to meet the demands of industrialisation, modernisation, international integration and the learning needs of the people (Khanh & Kham, 2013; *VOV News* 2017). So, what are the major obstacles and challenges causing the underperformance of education in Vietnam? Do the current waves of globalisation offer any opportunities for the country to reform and modernise its education system? How has the Vietnamese approach to education reforms changed under the forces of international integration? And to what degree does the approach fit into the national development strategies and regional integration needs?

This chapter first outlines a few defining opportunities and challenges – both internal and external – confronting Vietnam during its current educational reforms. The chapter then identifies and examines Vietnam's approach to education reforms in recent years which is characterised by three major features – (i) outward-looking education, (ii) socialisation of education, and (iii) job-hunting first approach. The chapter concludes with a discussion of how education reform can be guided by practice-based approaches in developing countries such as Vietnam.

Transforming education in Vietnam: opportunities and challenges

Under the forces of globalisation and deeper regional integration, Vietnamese education appears more receptive and innovative to global education. Internal institutional and policy reforms, years-long sustained economic growth followed by an expanding middle-class and increased inflows of foreign investment all have resulted in dynamic opportunities to promote innovative education in Vietnam.

Since the country is entering an epoch of modernisation and industrialisation, global economic integration creates a constant stream of new, highly-specialised jobs. To create a workforce capable of supporting a different type of economy, improving the education system is absolutely necessary (Ness & Lin, 2015). Faced with such a context, Vietnam has propelled education to the top of the government's agenda, while the amount of skilled labour is also deemed by the government to be 'the vital weapon' to increasing national competitiveness and bolstering investment. Thus, education is one of the nation's top priorities, enabling the MOET to allocate greater central funds to facilitate education innovation efforts.

Additionally, the varying educational demands and diverse-job hunting needs of the booming middle-class population and young labour market are driving practice-based education reforms in Vietnam. Due to continuously strong economic growth and rising disposable incomes, many families – from rural to urban areas – are now able to afford private tutoring and are looking outside of the limited education services in state-funded schools. Furthermore, in recent years investors and civil society groups have played their part in education development through their pro-active investment and engagement in semi-public and private education sectors (AUCV, 2013; Xuan Trung, 2014; The Nam & Manh Thang, 2016). Given the proliferation of these non-public educational institutions, general education has witnessed profound changes, turning from a highly centralised educational institution to a socialised one where the pressure to compete against one another has led both public and private schools to improve their academic quality, governance capacity and facilities (Ness & Lin, 2015; Horn, 2014).

Overseas Vietnamese or 'Viet Kieu' have long been supported in terms of financial and human resources within their own community. Vietnamese diasporas predominantly living in the US, and other countries that experience the world's best educations such as Australia, France, Canada and Russia, have been involved. As part of the national conciliation policy, the communist government has opened the door wider to millions of Vietnamese exiles to return and send money back to Vietnam. According to data released by the World Bank and the International Monetary Fund, since 2000 onward some US$6–8 billion (about 8 per cent of the country's GDP) has been sent to Vietnam each year to sponsor education and healthcare, expand businesses and support extended families. These capital flows have significantly contributed to improving the standard of living, and the remitters now tend to devote their funds to education development through multimillion-dollar school-building projects and to sustaining private universities whose owners primarily are, or backed by, 'Viet Kieu' (*Viet Nam News*, 2015).

Not only do the overseas Vietnamese communities contribute to the country's development through remittance flows, but many Vietnamese refugees in the 1970s and their children are now recognised professors at world-class institutes, and have been involved in instruction and management tasks at universities in Vietnam. Some have also played key roles in connecting the world's think tanks and educators to those in Vietnam, while others have engaged in the country's educational policymaking (Vietnam News Agency, 2016; Wang & Ji, 2016).

Another positive prospect is the emerging education trends. An 'open door' approach means an inevitable reception to the tsunami of culture and education, among other things, pervaded from the western societies. The increasing diffusion and penetration of Western-style education systems/liberal education from the West into less-developed countries/ postcolonial societies, on the one hand, is likely to bring more opportunities than threats – even though the threats, in turn, may challenge, and thus induce, the local Vietnamese education system to change and develop. On the other hand, regular exchange and international collaboration in today's globalised education appear to be narrowing academic gaps between different education systems around the world.

In this respect, Vietnam is a natural hotspot for education integration. Like Singapore, Vietnam's geographic position uniquely motivates it to support adequate approaches to education development as both a labour-intensive economy and as a potentially dynamic market for international education-related services. Increasingly, wealthy families in major cities such as Hanoi and Ho Chi Minh City and billionaire farmers in rural areas are seeking to enhance their children's education so that they may obtain skills which meet the ever-increasing demands of the labour market. Thus, foreign universities and private schools with state-of-the-art facilities and world-class teaching quality are being seen as the most effective means of obtaining a leg up for their children, especially when the competition for places at top educational institutions grows ever fiercer (Llewellyn-Jones, 2016; Stanley, 2017).

In recent years, foreign direct investment in education has begun to increase in Vietnam. Most foreign investors in education are keen on investment in the language prep market, prep schools and joint-venture universities, etc. At the higher education level, Vietnam also has established four universities of excellence and four other 100-per cent foreign-owned universities. Meanwhile, joint-venture international schools at general education level have started to emerge since 2005. Outstanding competitors include Vietnamese-American School system, Asia Pacific College system and Vietnam-Australia International School system, etc. with tens of centres in big cities. Given the traditional notion that non-public schools are less reputable than those state-run, the private education providers have employed long-term investment strategies to attract students and the Vietnamese public. Their strategies appear different from one another, but basically emphasise learning and teaching quality and more focus on having strong teacher training systems, comfortable learning environments with enough modern facilities and low student-teacher ratios.

Vietnam education reforms: unsettled challenges

Vietnam has made remarkable progress in education reforms since the first official review of the general education curricula in 2000. After years of piloting, debate and modifications, the new stream-lined curricula and textbooks have been introduced since the academic year 2006–2007 and used for teaching and learning at all grades at the general education level. Since then, several more reviews have been undertaken by the MOET to revise the curricula and textbooks. The 2000 review of the general education curricula was to meet the objectives and requirements of educational content and methodology prescribed in the Education Law (World Bank, n.d.). By 2016, the reviews had achieved encouraging outcomes. At its first participation in the OECD-organised Programme for International Student Assessment (PISA) in 2012, Vietnam's 15 year-olds ranked seventeenth in maths and nineteenth in reading, performing on a par with students in Germany and Austria and better than those in two-thirds of the participating countries. In the 2015 PISA test results, Vietnam was ranked eighth out of 72 countries in terms of its scientific performance.

During the period 2006–2015, the majority of tertiary institutions, particularly public universities, successfully refined the division of training programmes into two categories: research and career application. Furthermore, many tertiary institutions also shifted from old-fashioned year-based courses to a credit-based training system while institutional autonomy and accountability of all aspects have increasingly improved. The MOET has also finalised the higher education quality assurance and accreditation system, and further reached agreements of mutual recognition of qualifications with 16 countries.

Regarding equality in opportunities for access to education, the Vietnamese government has offered greater learning conditions for everyone to guarantee their learning and professional development rights. The government ensures this principle of social equity in education by strictly implementing policies on social subsidies, scholarships and fee exemptions or reductions with priorities given to ethnic minorities, children from socio-economically disadvantaged families, and people with disabilities. The government also provides financial support so that non-public schools will be able to offer fee exemption/reduction policies to their disadvantaged students (World Bank, n.d.; *PetroTimes*, 2016). Despite these signs of progress, Vietnam, however, needs a further strengthening of education reforms, especially at secondary and tertiary levels, due to domestic obstacles and emerging challenges.

At the pre-tertiary education level, Vietnam's general education curricula are still strongly influenced by the conventional approach of teaching mandatory subjects (e.g. literature, history, geography, mathematics, physics, chemistry) within their respective disciplines rather than through interdisciplinary approaches. This is because curriculum specialists/authors tend to impose the learning path and methods that they have previously experienced. Educational policy planners are also receiving criticism for their overdependence on ambitious yet unrealistic educational philosophies/models while paying little attention to, or without consulting, those directly affected – the teachers, the students and their parents. The current limitations of school management capacities and teacher qualifications appear unable to meet the requirements of any revised curriculum (World Bank, n.d.; Horn, 2014; Thuy Linh, 2015).

In the higher education sector, Vietnam's higher education retains ill-suited courses which lack responsiveness to the rapid changes of society and the economy. It is difficult to overstate the seriousness of the challenges confronting Vietnam's tertiary education, but the existing lack of even a single university appearing in any of the recognised league tables of the leading Asian universities appears to affirm the sentiment that, 'Vietnamese universities lag far behind even their undistinguished Southeast Asian neighbors' (Vallely & Wilkinson, 2008). In addition, the poor quality of tertiary education – both undergraduate and post-graduate – adds another flaw to higher education development in Vietnam (Tuoi Tre News, 2014).

From the standpoints of educators and experts, the set-back of Vietnam's higher education has long been caused by internal factors – historical legacy and poor governance. Particularly, due to decades-long tragic and devastating wars throughout its modern history, the war-torn country had missed the waves of institutional innovation in higher education brought from the West during the twentieth century, when many regional leading tertiary institutions were established in Singapore, Thailand and the Malay World (Vallely & Wilkinson 2008; Woo & King, 2013). Furthermore, after Saigon's surrender in 1975, Hanoi eliminated all educational advisors trained in the West and quickly became a carbon copy of its communist mentors – the Soviet Union and the People's Republic of China – by employing a subsidised education system characterised by teacher-delivered lectures, or factory-model schools. The bitter legacy of this educational style still runs deep in lecture halls across Vietnam, challenging the country's transformative education process.

Vallely and Wilkinson (2008) in their report entitled 'Vietnamese Higher Education: Crisis and Response' conclude that profound governance failure is the most immediate reason causing the poor quality of Vietnamese higher education. Many agree that the tertiary institutions lack autonomy since the education system is highly centralised and completely controlled by the communist-ruled government. Additionally, a prevailing lack of transparency in personal affairs, accountability and meaningful international connections in most of the universities are other prime culprits and intractable hurdles, denying the universities and institutes the incentive to compete or innovate. These problems also undermine intellectual dynamism and academic freedom and prevent Vietnam's post-secondary institutions from doing well on global education rankings.

Vietnam's approach to education reforms

With an outdated education system set to shoulder the increasingly unbearable impacts of the above-mentioned historical legacy and poor governance, Vietnam has persistently made many modifications to curricula, educational perceptions and textbooks over the past 18 years. However, the outcomes of those reforms are allegedly modest, challenging any educational modernisation attempts. To that end, in November 2015 the MOET once again announced plans to launch ambitious schemes on education reform (Thuy Linh, 2015).

In April 2017, a new curriculum scheduled to be in place in the academic year of 2018–2019 was introduced by reputable educators, curriculum planners and educational policymakers. The MOET-championed an 80 million-dollar programme, that 'seeks to thoroughly transform the country's national education by introducing a shift in educators' outlook on pedagogy' and 'transforms general education from a system that provides knowledge into one that develops capacity' (Thuy Linh, 2015; Tuoi Tre News, 2017).

In order for Vietnam's tertiary education to keep up with international standards, the new leadership in MOET has urged state and private post-secondary institutions to innovate and modernise in their own ways through any legal means available (*Dan Tri News*, 2016; Phuong Linh, 2017). The minister also commits to governance reforms, including greater academic freedom and financial autonomy and more merit-based selection mechanisms in universities. As part of the effort to achieve regional and international integration in the education sector, the ministry is willing to actively participate in activities led by regional and international organisations such as the United Nations Educational, Scientific and Cultural Organization (UNESCO), the United Nations International Children's Emergency Fund (UNICEF), the United Nations Development Programme (UNDP) and the Southeast Asian Ministers of Education Organization (SEAMEO). In addition, it further calls upon foreign education investors and the world's leading universities to establish institutions/training programmes in Vietnam, either independently or, most commonly, in partnership with Vietnamese public institutions, and other direct investments in high-quality infrastructure, physical facilities and equipment (World Bank, n.d.; Vallely & Wilkinson, 2008).

Given the anticipated challenges in the coming years, Vietnam apparently has been trying to triumph over its existing adversities to improve the standards of its education system. The country's enduring endeavours have constituted what may be called 'the Vietnamese way of education for sustainability', which overall aims to translate the current educational style into a transformative learning process that equips students with new knowledge, soft skills and perspectives to achieve economic prosperity at the local and international levels, pursue lifelong learning, further professional development and become responsible citizens.

The Vietnamese way is characterised by education policies and strategies that incline the national education system to be more competitive, open-minded and modernised. The following sub-sections outline three strategies that shape the Vietnamese approach to modernising education, namely: (i) outward-looking education, (ii) socialisation of education, and (iii) 'job-hunting first approach'. These are also the core elements of the Vietnamese way of Education for Sustainability, in line with the UNESCO Sustainable Development Goal (SDG) 4, on providing quality education.

Outward-looking education

Objectively speaking, the Vietnamese academic environment has long been very inward looking and has not matched any international standard. Thus, under the forces of deterritorialisation, followed by regional integration in education, the recent education reform schemes appear to deal with the existing weaknesses and mismatches. The common objectives of such reform measures eventually focus on developing an outward-looking education system.

Specifically, in the general education sub-sector, the reform plans aim to reduce pressure on students, focusing on developing integrated elementary and lower secondary education (grades 1–9), and a differential and elective upper secondary education (grades 10–12). At the pre-high school level, students are equipped with general knowledge on basic disciplines – Vietnamese literature, maths, arts and music, geography and history, and technology – to form and develop various qualities, including patriotism, charitableness, diligence and discipline, honesty and responsibility. At the high school level, the 10th grade is identified as a career orientation grade with 11 compulsory courses. These courses are reduced to six in grades 11–12 so that students can choose electives that complement their career choices. High school students are also expected to possess and develop core competencies such as self-control, self-learning, communication, collaboration, problem solving and creativity. To meet these goals, many high schools have applied active-learning methods with teachers in the role of a coordinator, instructing and encouraging students to take part in educational and capability-developing activities. These activities are held inside and outside of classrooms as theoretical lectures, experiments, role plays, research projects, conferences and picnics, as well as collective and community service activities. It is hoped these will help students live and work in the future.

As for higher education, the degree of autonomy accorded to universities has gradually increased. Public universities are given autonomy in all aspects – training, research, international cooperation, organisation, personnel management and financial affairs. The MOET has expressed a willingness to decentralise the education system by reducing the role of the government in university operations and academic collaborations, empowering post-secondary institutions to work with Western education institutions.

Looking externally, the government acknowledges that overseas training of graduates provides an efficient way of acquiring advanced knowledge and research and development experience for Vietnam. To that end, it has offered up to 10,000 government scholarships for overseas doctoral study between 2014 and 2020 to tertiary and research institute staff through numerous scholarship schemes, notably the well-known 911 Project and other state-funded scholarship and fellowship projects such as the 599, 322, 165 and Mekong 1000 projects. The most popular education destinations for selected scholarship recipients are the US, UK, Australia, China, South Korea, Germany, the Netherlands, Russia, Japan and Taiwan. It is worth observing that not only does the success of these projects and programmes lie in how many Western educated professionals return to work in Vietnamese tertiary institutions, but also in productive collaboration with overseas institutions (Hoa, 2016).

Furthermore, the current higher education reforms reveal that Vietnam plans to 'outsource' higher education to foreign universities, especially from the US, UK and Australia, by fostering and actively preparing promising proposals for cooperation with those stellar education systems. In addition, some Vietnamese universities have begun offering professionals who study abroad better working conditions and attractive incentives.

To address the underperforming education quality, the MOET issued legal grounds for quality assurance and enabled universities to use quality assessment tools offered by international independent agencies such as ABET and AUN-QA. By 2016, there were 35 advanced programmes in association with 23 foreign universities and 17 high-quality undergraduate engineering programmes established in ten leading universities in Vietnam.

Most of the undergraduate curricula in public universities have been revised and updated to match lectures with employers' demands and global mainstream knowledge evolution. Concepts-intensive modules have been replaced with practice-based coursework which emphasise student capacity-building, practice-based vocational skills/professional skills and lifelong learning skills. Each year, thousands of undergraduates are selected to attend advanced programmes and international exchange courses in universities outside Vietnam. This enhanced student mobility is an integral part of Vietnamese universities' internationalisation efforts.

Socialisation of education

Vietnam implemented policies of educational socialisation in the early 1990s with cautious steps. Since then, permission has been granted to open private kindergartens, and semi-public and private schools at all levels, including the post-secondary level. In 1994, the number of publicly-funded schools was limited: five universities and 12 lower and upper secondary schools mostly located in major cities. By 2017, the number of private universities and pre-tertiary schools had reached 60 and 37 respectively, with hundreds of centres across Vietnam, handling over 13 per cent of undergraduate students and approximately 4 per cent of lower and upper secondary students in the school year 2015–2016 (Nguyen Dung, 2017).

Some joint-venture or 100-per cent foreign-invested secondary schools have provided bilingual curricula (English and Vietnamese) while foreign universities such as the Vietnam-German University and Royal Melbourne Institute of Technology (RMIT), for example, offer English-speaking programmes. Though non-public education services do not come cheap in Vietnam, many private and foreign institutions usually provide international-quality education at about one-third or half the cost of studying overseas. Thus, those institutions have become an option for the increasing number of students who either have the ability to pay or are fortunate to win a scholarship. The success story of some private and foreign universities like Financing and Promoting Technology (FPT) University and RMIT are expected to become a catalyst that creates a competition with the state-run universities.

In this respect, the socialisation of education, on the one hand, contributes to improving and diversifying the national education system. On the other hand, the policy represents the government's commitment to surrender its monopoly on shaping the minds of young Vietnamese and appeal to both domestic and foreign private investors to help improve its education system. The swiftly expanding private and foreign-funded education sub-sectors have the potential to trigger a quiet yet profound race on innovation in terms of curriculum, training approaches, methodology, international collaboration and quality, between public and non-public education institutions at all levels in the near to medium term.

'Job-hunting first' approach

This approach appears to address existing mismatches between Vietnam's higher education system and the high rate of unemployment upon graduation (Kim Ngan, 2012; Minh Duong, 2016). The current education reform plan made public in April 2017 is an example of this approach. For years, the Vietnamese educational policymakers and educators have resolved to apply career orientations in high school activities to help students finalise their choice of specialisation that they will pursue at higher education institutions. Meanwhile, in universities, undergraduates are encouraged to take part in practice-based modules so as to acquire practical experience upon entering the workplace. Multimillion-dollar technology transfer projects have been approved by the government to improve school facilities and university labs. Additionally, more vocational education centres and universities now tend to move closer to employers, companies and recruiters in order to understand and keep themselves well-informed with the latest changes in the labour market. By doing so, education institutions can modify their admission requirements and training programmes to ensure that their students' qualities fit neatly with society's demand.

The Vietnamese people also acknowledge that low English proficiency skills are one of the biggest obstacles that prevent students from approaching international high-quality training, professional development and high-paid job opportunities. In fact, many Vietnamese graduates who have good professional knowledge are often unsuccessful in job-hunting due to limited foreign language skills. They also tend to lack the negotiation skills needed when dealing with employers, especially those from abroad. Thus, to address this shortcoming, the government has approved an ambitious plan called 'the National Foreign Languages Project 2020'. One of the objectives of this US$450 million project is to enable Vietnamese youth to be more confident in communication, enhancing their chances to study and work in an integrated and multi-cultural environment. Vietnamese parents, especially in urban areas, now enrol their children in English classes at very early ages. The Vietnamese people are now spending much more time and money on learning the English language than many families in more developed but not English-speaking countries (*Vietnamnet Daily*, 2016).

Concluding remarks

Decades of relentless education reforms and modifications since Doi Moi have witnessed some remarkable progress and improvement in education quality and integration in Vietnam. The early benefits of Vietnamese education reforms result from dynamic opportunities – both domestic and foreign – while 'made-in-Vietnam' factors such as poor governance and historical legacy are driving factors causing poor education quality at all levels and challenge the future development of the education system.

However, by observing the measures and policies undertaken by the Vietnamese government in the past and current reforms, this chapter identifies three strategies that shape the Vietnamese way of education for sustainability. These include the outward-looking policy which stresses international cooperation, the educational socialisation policy empowering social resources to get engaged in the government's education reforms, and the 'job-hunting first' approach that emphasises the career-oriented training. These are integral elements to ensure the country's education modernisation which in turn produces qualified 'products' – skilled labours – to satisfy the requirements of highly-specialised jobs in today's knowledge-based economy.

As a developing country set to face increasingly fierce challenges posed by the forces of globalisation, Vietnam apparently wants to overcome adversity and sustain socio-economic development by boosting its attempts at education reforms. This is clearly a prudent policy alternative for self-help economic development in poor countries such as Vietnam where the pressure to compete internationally always requires both an increase in educational efficiency and an investment environment that encourages skilled workers and the acquisition of new technology.

Accordingly, education quality is being improved in line with SDG 4, and there is a rapid shift away from liberal-arts education towards lucrative career-oriented education. Educational development in this environment is no longer constrained by how fast the liberal-arts colleges can progress; instead, Vietnam is now able to target external job markets by providing competency-based education that appears to fit neatly into the international qualification standards. In this respect, the chapter sheds new light on an approach to Education for Sustainability in developing countries where the transformative learning process now aims to equip learners with new practice-based knowledge and skills to achieve economic prosperity in a globally-competitive environment. By proposing the concept of 'job-hunting first approach' as the key to understanding contemporary education reforms launched by Vietnam, the chapter also elaborated on how education reform can be guided by practice-based approaches in developing countries.

References

Association of Universities and Colleges in Vietnam (AUCV). (2013). *Report on 20 Years of Development Models of Higher Education Outside the Public in Vietnam*. Retrieved from http://hiephoidaihoccaodangvn.vn/20-nam-phat-trien-mo-hinh-dai-hoc-ngoai-cong-lap/.

Daniel, N., & Chia-Ling, L. (2013). *International Education: An Encyclopedia of Contemporary Issues and Systems (2-Volume Set)*. London: Routledge.

Dan Tri News. (2016, 1 September). Bộ trưởng Phùng Xuân Nhạ đặt 9 nhiệm vụ, 5 giải pháp để nâng kỷ cương, chất lượng giáo dục. Retrieved from http://dantri.com.vn/giao-duc-khuyen-hoc/bo-truong-phung-xuan-nha-dat-9-nhiem-vu-5-giai-phap-de-nang-ky-cuong-chat-luong-giao-duc-20160901072010179.htm.

Hoa, T. V. (2016, 21 April). Grading Vietnam's higher education reforms. *East Asia Forum*. Retrieved from www.eastasiaforum.org/2016/04/21/grading-vietnams-higher-education-reforms/.

Horn, M. (2014, 24 February). Visits to Vietnam's schools shed light on opportunities for innovation. *Forbes*. Retrieved from www.forbes.com/sites/michaelhorn/2014/02/24/visits-to-vietnams-schools-shed-light-on-opportunities-for-innovation.

Khanh, N. V., & Kham, T. V. (2013). Opportunities and challenges for Vietnam's higher education in the globalisational process: Experiences from Hanoi University of Social Sciences and Humanities. *US–China Education Review B, 3*(9), 680–689.

Kim Ngan. (2012, 20 September). 63% sinh viên thất nghiệp, chất lượng giáo dục có vấn đề? *Giao Duc News*. Retrieved from http://giaoduc.net.vn/Giao-duc-24h/63-sinh-vien-that-nghiep-chat-luong-giao-duc-co-van-de-post88908.gd.

Llewellyn-Jones, G. (2016, 21 April). Industry spotlight: Vietnam's growing appetite for education. *Vietnam Briefing*. Retrieved from www.vietnam-briefing.com/news/industry-spotlight-vietnams-growing-appetite-education.html/.

Minh Duong. (2016, 3 March). 90% học sinh muốn học đại học, cao đẳng để rồi... thất nghiệp. *VOV News*. Retrieved from http://vov.vn/xa-hoi/giao-duc/90-hoc-sinh-muon-hoc-dai-hoc-cao-dang-de-roi-that-nghiep-491408.vov.

Ness, D., & Lin, C. L. (2015). *International Education: An Encyclopedia of Contemporary Issues and Systems*. Abingdon: Routledge.

Ngoc, D. T. (2016, 12 November). Chấm dứt việc đọc – chép, phong trào rầm rộ vài bữa rồi thôi! *Giao Duc News*. Retrieved from http://giaoduc.net.vn/Giao-duc-24h/Cham-dut-viec-doc-chep-phong-trao-ram-ro-vai-bua-roi-thoi-post172053.gd.

Nguyen Dung. (2017, 15 April). Đại học ngoài công lập: Đòi bình đẳng với trường công. *Tien Phong News*. Retrieved from www.tienphong.vn/giao-duc/dai-hoc-ngoai-cong-lap-doi-binh-dang-voi-truong-cong-1140331.tpo.

Nhan Dan Online. (2018). Optimising the advantages of Vietnam's golden population structure. Retrieved from https://en.nhandan.org.vn/society/item/6795602-optimising-the-advantages-of-vietnam%E2%80%99s-golden-population-structure.html.

PetroTimes. (2016, 27 September). Hơn 3,3 triệu lượt học sinh, sinh viên nghèo được vay vốn tín dụng ưu đãi. Retrieved from http://petrotimes.vn/hon-33-trieu-luot-hoc-sinh-sinh-vien-ngheo-duoc-vay-von-tin-dung-uu-dai-485476.html.

Phuong Linh. (2017, 14 April). Bộ trưởng Phùng Xuân Nhạ trả lời câu hỏi về cách mạng công nghiệp 4.0. *Giao Duc News*. Retrieved from http://giaoduc.net.vn/Giao-duc-24h/Bo-truong-Phung-Xuan-Nha-tra-loi-cau-hoi-ve-cach-mang-cong-nghiep-40-post175808.gd.

Stanley, S. (2017, 15 April). The good, the bad and the ugly: Vietnam's booming after-school tutoring business. *VNExpress Newspaper*. Retrieved from http://e.vnexpress.net/news/business/the-good-the-bad-and-the-ugly-vietnam-s-booming-after-school-tutoring-business-3570388.html.

The Nam and Manh Thang. (2016, 15 March). Khởi công xây dựng 3 phòng học Dân trí tại Yên Bái. *Dan Tri News*. Retrieved from http://dantri.com.vn/tam-long-nhan-ai/khoi-cong-xay-dung-3-phong-hoc-dan-tri-tai-yen-bai-20160315071005502.htm.

Thuy Linh. (2015, 11 November). Vietnam's ambitious education reform plans come in for praise. *Thanh Nien News*. Retrieved from www.thanhniennews.com/education-youth/vietnams-ambitious-education-reform-plans-come-in-for-praise-53391.html.

Tuoi Tre News. (2014, 9 October). French Prof sends letter to Vietnam education minister over doctoral training flaws. Retrieved from http://tuoitrenews.vn/education/22317/french-prof-sends-letter-to-vietnam-education-minister-over-doctoral-training-flaws.

Tuoi Tre News. (2017, 13 April). Vietnam to scrap national high school exam in bid to transform education system. Retrieved from http://tuoitrenews.vn/education/40505/vietnam-to-scrap-national-high-school-exam-in-bid-to-transform-education-system.

Vallely, T., & Wilkinson, B. (2008). *Vietnamese Higher Education: Crisis and Response*. Cambridge, MA: Havard Kennedy School, Ash Institute.

Viet, L. C. L. (2009). Education Reform in Lower Secondary Education in Vietnam. In Hirosato, Y, & Kitamura, Y. (Eds.), *The Political Economy of Educational Reforms and Capacity Development in Southeast Asia*. doi:10.1007/978-1-4020-9377-7_13.

Vietnamnet Daily. (2016, 21 June). Poor English skills challenge Vietnamese students. Retrieved from http://english.vietnamnet.vn/fms/education/159027/poor-english-skills-challenge-vietnamese-students.html.

Viet Nam News. (2015, 25 December). Overseas Vietnamese retain a role in national development. Retrieved from http://vietnamnews.vn/opinion/280360/overseas-vietnamese-retain-a-role-in-national-development.html.

Vietnam News Agency. (2016, 9 September). La Professeure Lê Kim Ngoc reçoit la Légion d'honneur. Retrieved from http://fr.vietnamplus.vn/la-professeure-le-kim-ngoc-recoit-la-legion-dhonneur/79766.vnp.

VOV News. (2017, 4 July). Adapting higher education in Vietnam to globalization. Retrieved from http://english.vov.vn/society/adapting-higher-education-in-vietnam-to-globalization-347133.vov.

Wang, D., & Ji, L. (2016, December). Vietnamese mathematician Ngô Báo Châu: from A mathematical Olympiad medallist to a fields medallist. *Asia Pacific Mathematics Newsletter,* 6(2). Retrieved from www.asiapacific-mathnews.com/01/0102 /0025_0030.html.

Woo, P. S., & King, V. T. (Eds.). (2013). *The Historical Construction of Southeast Asian Studies: Korea and Beyond*. Singapore: Institute of Southeast Asian Studies.

World Bank. (n.d.). *Report on Education in Vietnam: Development History, Challenges and Solutions*. Retrieved from http://siteresources.worldbank.org/EDUCATION/Resources/278200-1121703274255/1439264-1153425508901/Education_Vietnam_Development.pdf.

Xuan Trung. (2014, 18 March). Chỉ có xã hội hóa mới mang lại bộ mặt mới cho giáo dục. *Giao Duc News*. Retrieved from http://giaoduc.net.vn/Giao-duc-24h/Chi-co-xa-hoi-hoa-moi-mang-lai-bo-mat-moi-cho-giao-duc-post141584.gd.

12 Experiences and lessons

An international training programme on Education for Sustainability in the context of Chinese formal education

Qi Zhang

Introduction

The Swedish International Development Agency (SIDA)-supported 'Environmental Education (EE) and Education for Sustainability (ESD) in Formal Education' was run exclusively for Chinese participants from 2004 to 2009. This International Training Programme (ITP) on EE and ESD was established at a time when the Chinese government had increasingly realised the negative consequences of rapid economic growth. As a response to the environmental, economic and social problems faced, the national development strategy used sustainable development or 'harmonious development' as a guiding principle, trying to mitigate its negative impacts on the ecosystem.

Accordingly, in the formal education system, the Ministry of Education (MOE) has initiated radical curriculum reform since the middle of the 1990s (Ministry of Education, 2003). The revised curriculum was decentralised and outcomes-based. It gave more responsibility and freedom to local educational authorities, individual schools and subject teachers than ever before. This allowed them opportunities to develop their own curricula and work towards the expected outcomes. In October 2003, the MOE released the National Environmental Education Implementation Guidelines for students in grades 1–12 after wide consultation with international and national EE researchers and practitioners (Qiaoling Wang, 2015). This document included objectives, principles and examples on how EE could be integrated into curriculum mainly in a cross-disciplinary way. This situation was a window of opportunity for EE and ESD. It would be seen as a useful medium for propelling the curriculum reform. The ideas and approaches adopted by the training programme could give specific examples of quality education that were significant for the curriculum reform in teacher training, curriculum development and policy implementation.

Process of implementation

Partnership

Administrated by NIRAS AB (formerly RAMBOLL Natura AB), a Swedish consultancy company, the training programme was built on unique partnerships. Within Sweden, the SIDA Advisory Committee on Environmental Education played a significant role in the initial development of the programme. A number of Swedish universities, schools, teachers and organisations were involved in developing and implementing the programme. Internationally, Southern African Development Community–Regional Environmental Education Program (SADC-REEP) and the Environmental Education Centre of the East China Normal University (EEC ECNU) in Shanghai contributed an international perspective and an informed position on the curriculum reform in China. Within China, the Department of Basic Education, MOE and the Centre for Environmental Education and Communication, Ministry of Environmental Protection (CEEC MOEP) provided institutional supports. The Department of Curriculum and Instruction in ECNU, one of the most leading institutions in shaping the national curriculum standards, and a group of green schools throughout the country, has added to the relevance and quality of the programme significantly.

Objectives

The main objectives of the training programme were to produce an opportunity to:

- Exchange knowledge and experiences in EE and ESD processes in the context of curriculum reform in China and
- Use the insights gained to enhance and develop EE and ESD in the participants' work contexts.

The specific objectives of the programme included:

- To understand environmental and development issues and risks from a range of perspectives at global, regional and local level;
- To increase knowledge of current trends and international initiatives in EE and ESD;
- To develop a broad understanding of the institutional framework and the different structures needed for effective implementation of EE and ESD;
- To explore ways of integrating EE and ESD processes into formal school curricula and teacher training especially in the light of the National Environmental Education Implementation Guidelines;

- To increase knowledge of different methods and processes in EE and ESD, many of which are relevant in the context of the curriculum reform in China;
- To locate the learning in the working contexts of the participants and
- To develop the link between the EE and ESD policy and the working context of the participants.

Participants

The ITP created a meeting place for more than 180 (around 30 for each year) EE and ESD professionals from all over China between 2004 and 2009 (see Figure 12.1).

The Department of Basic Education, Ministry of Education and the Centre for Environmental Education and Communication, Ministry of Environmental Protection recommended most of the participants. Priority was given to senior school managers, school inspectors, curriculum developers and policy makers. All of them were based in or directly supporting the formal education system and had multiple potentials to implement change (see Figures 12.2 and 12.3).

Course elements

In this programme, three key themes were explored. These were (a) issues of environment and sustainability; (b) EE and ESD methods and processes; and (c) policy, curriculum and resource materials as supporting structures. The focus was explicitly on the inter-relationships among them. For example, the understanding of environmental issues as social constructs required education processes that moved beyond knowledge delivery and awareness rising to critical processes of contextual action taking. Furthermore the links between these contextual education processes that engaged with learners' lives had close relation to the kind of curriculum reform undertaken in China.

Starting from an institutional audit, the ITP provided three weeks of intensive training in Sweden, mainly on the theories and practices of EE and ESD from international perspectives through presentation, school visits, evaluation, peer discussion and self-learning. Then the participants went back to their home institution to share the insights gained with their colleagues through institutional workshops or seminars and to formulate their own project outlines. In part 4, the change ideas opened up earlier were explored in depth with support from the Chinese resource persons in the regional component. By the end of the two weeks in ECNU, the participants would finalise their own contextualised change projects (see Figure 12.4).

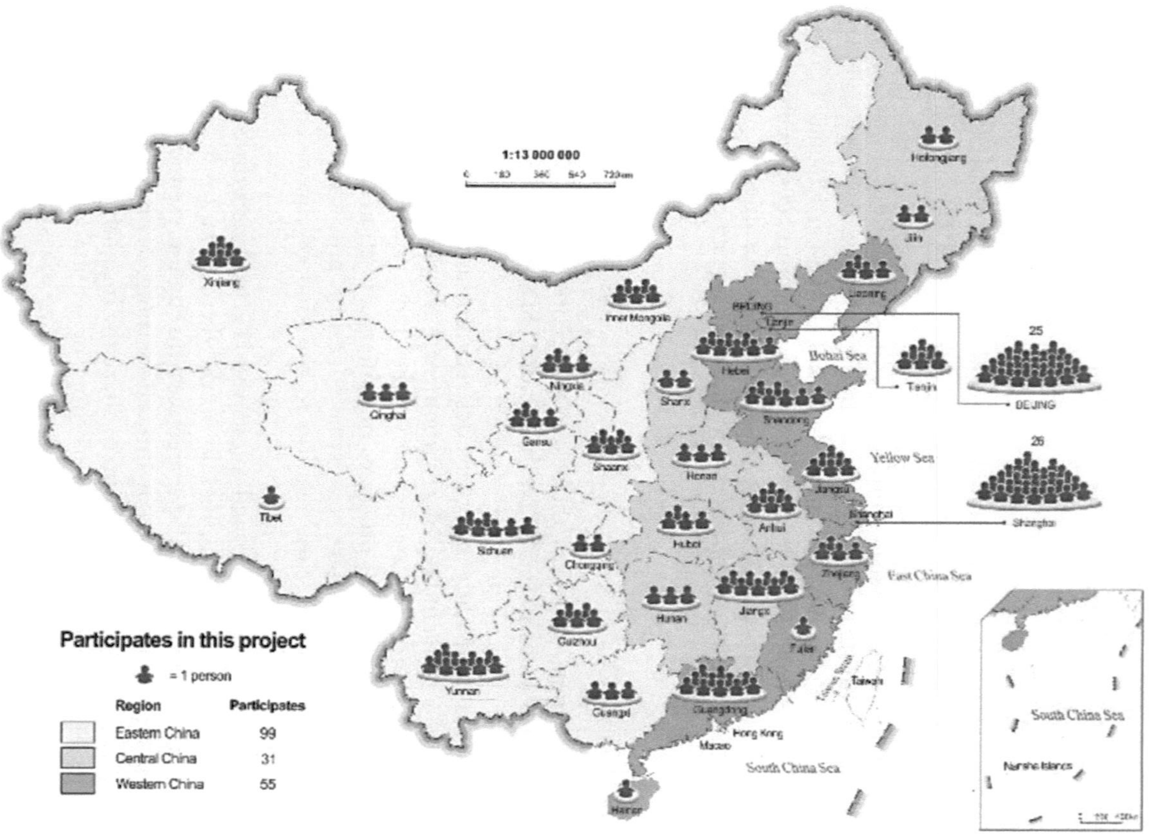

Figure 12.1 Participants by province (2004–2009).

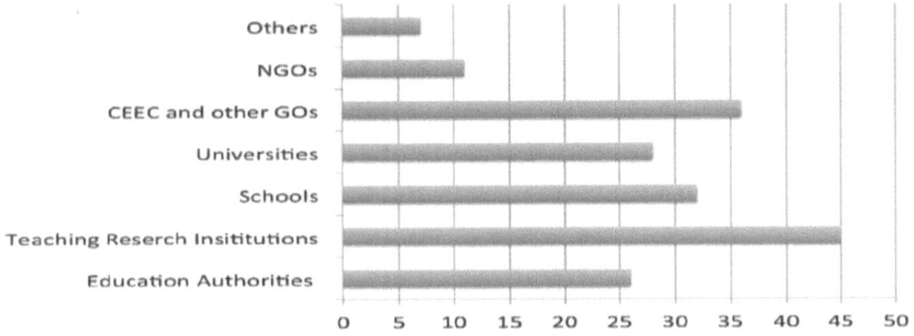

Figure 12.2 Participants by sector (2004–2009).

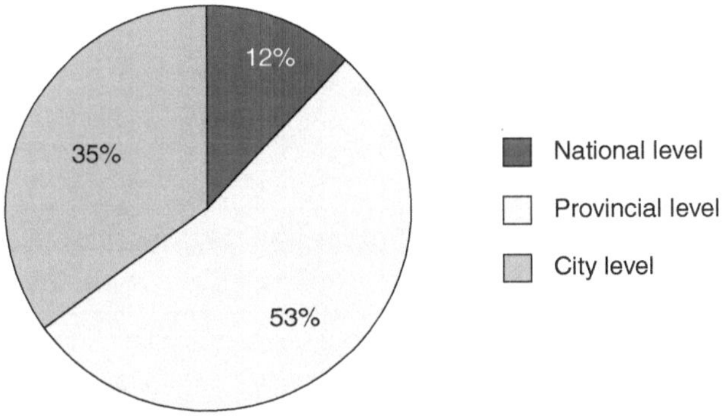

Figure 12.3 Participants' organization by level (2004–2009).

Change projects implementation

The learning from the programme was put into practice in the participants' own work contexts. From 2004 to 2009, 174 change projects were produced and some have resulted in great institutional changes from national levels to school levels. Below are some examples of the types of impact of the different projects (see Table 12.1).

Reflection on the outcomes

Experiences

Strong partnership

Internationally, Sweden has been an active player in the implementation of the UN Decade (2005–2014) on ESD. There were prominent research groups at Swedish universities such as Uppsala, Malmö and Lund doing research on ESD. The development of this ITP also had close links to

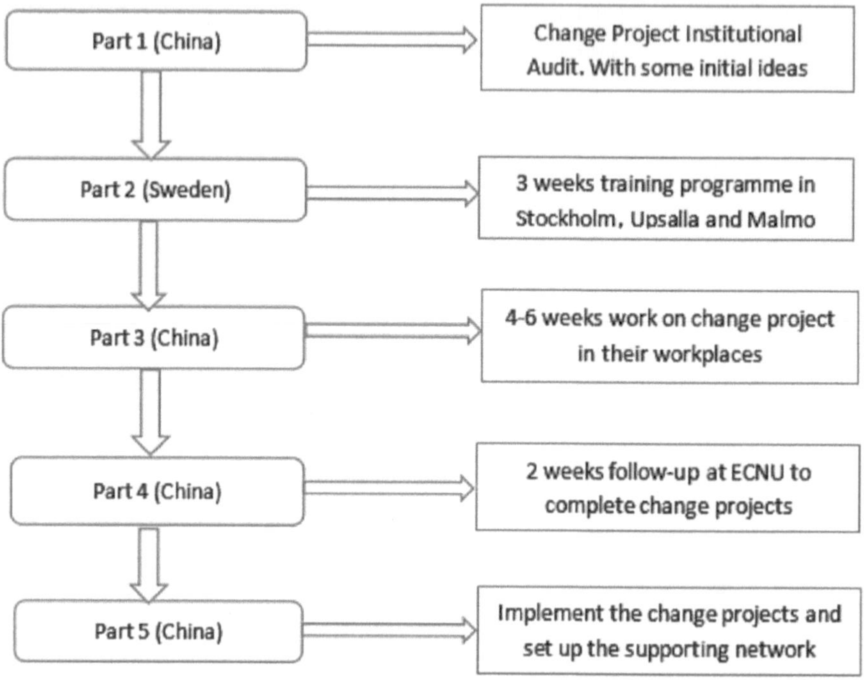

Figure 12.4 Programme framework.

another SIDA funded ITP with the same name, which was offered to Southern African and Asian participants in cooperation with the Environmental Learning Research Centre at Rhodes University, South Africa and the Centre for Environmental Education (CEE) in India since 2002. Both regional partners were leading organisations in EE and ESD research and had many experiences at the global and regional level for many years. The organisers of the two programmes drew on each other's experiences and incorporated and adapted positive changes into respective programmes. Within China, the programme has made the most use of ECNU's strength in teacher education, responded to the specific challenges of EE and ESD effectively, and communicated through participants and resource persons from different education and environmental protection organisations.

Responsive content

China had distinctive environmental, social, economic and political backgrounds compared with Sweden, South Africa and India. Course materials were therefore reoriented to reflect the emerging environmental and educational issues in China, for instance rapid urbanisation and biodiversity loss, trans-boundary toxic e-waste flow and consumerism, toxic haze and urban

Table 12.1 Example of Change Projects

Change Project	*Participant*
Policy proposal *Implementation Guideline on EE as a Local-based (Provincial level) curriculum in Compulsory Education (1–9),* officially released in 2008 and almost 10 million students were reached in Shandong Province	Mr Li Jiaqing, batch 2007, vice director of the Division of Basic Education, Shandong Provincial Education Department
Management practice *Guideline for Eco-School Initiatives,* adopted by CEEC, MOEP in place of Green School Program. Up to 2016, 335 schools had been awarded the Green Flags	Mrs Song Xuhong and Yang ke, batch 2009, director and program official of CEEC, MOEP
Management tool *Evaluation Packet for EE Mentoring Program,* used by 320 Roost & Shoots pilot schools in formal education	Ms Zhong Zhenxi, batch 2005, director of Roost & Shoots Shanghai Office (NGO)
School management, school-based curriculum, teachers training, teaching approaches, teaching-learning resource development etc. *Bilingual (Chinese & English) Course Plan on Environmental Protection in Shanghai Yan'an Junior Middle School.* It was a school-based optional course developed by an English teacher in cooperation with geography teachers	Mrs Shi Yinghua, batch 2009, English teacher of Shanghai Yan'an Junior Middle School

transportation. Participants' institutional audits were analysed and used to shape the programme schedule. Resource persons and mentors paid more attention in supporting the participants to adapt the international experiences to the Chinese context.

Interactive learning process

The national curriculum reform in China starting from the middle of the 1990s introduced students-centred learning in place of teachers-centred learning. The learning objectives were diversified in three dimensions: knowledge and skills, processes and methods, emotion, attitudes and values. This ITP valued *learning from doing, theory in practice and assessment as learning.* Active, cooperative and inquiry-based learning approaches were employed throughout the process. All participants were educators and were asked to contribute to panel discussions focused on their work and pre-course assignments. Peer reviews and critical comments from resource persons and mentors also enhanced the learning opportunities. In accordance with the new curriculum requirements, it was an excellent example of quality education. Participants' experience with the ITP empowered them to put the ideas and approaches promoted by the national curriculum reform into practice.

Supportive network

Many cases have shown that in struggling with the examination-oriented structures in formal education, some participants have developed their own EE and ESD programmes and expanded their own professional networks. One of the excellent examples was from *Water School China (*Shangri-la Institute for Sustainable Communities, 2014). It was a community-based ESD programme organised by an NGO, Shangri-la Institute for Sustainable Communities. Since 2008, to promote sustainable water resources management, more than 103 primary and secondary schools, 3,000 teachers, 197,000 students and more than 300,000 community members across China have been involved in the network on water education. In addition, 8 EE and ESD ITP alumni, played key roles in the programme development and implementation. They worked in five locations (Beijing, Shangri-la, Mianyang, Chongqing, Shanghai) and conducted activities, including teacher training, teaching and learning resource development and community engagement. As one of the outcomes, *Water-based EE Activity Package (Grade 1–12)* was published by China Environmental Science Press in April 2014. It was developed by a group of ITP alumni.

Lessons

Lack of following-up evaluation

During the training programme, the following evaluations on the contents, methodology and administration were carried out.

- Immediate feedback on a daily basis;
- Written weekly evaluations in Sweden and China (five weeks);
- Discussions and analysis as evaluation of the school visit and field visit in China;
- SIDA online evaluation by the end of Swedish component;
- SIDA online evaluation by the end of Chinese component.

The participants' feedbacks have been extremely helpful to improving the programme year by year. However, there was no follow-up evaluation on the change projects implementation. The quality of learning during the programme and impacts of change projects after the programme basically depended on the commitments of the participants and their organisations. Some were highly self-motivated since then and some were always silent.

Lack of academic research

Many participants described the most challenging parts of the programme as (i) Managing Change; (ii) How to apply learning from the course to work in China; and (iii) EE and ESD evaluation and assessment in formal education. This was useful information for the organisers. The areas where challenges were encountered were likely to be where the most effort was made. In fact, those challenges have not been responded to sufficiently. Even until this day there are quite a few research groups at Chinese universities doing theoretical research on EE and ESD like Uppsala, Malmö and Lund universities in Sweden, Rhodes University in South Africa and Centre for Environmental Education in India. The language issue makes it more difficult for Chinese EE and ESD practitioners to communicate with the international community.

This chapter has provided examples of how several programmes, with the continuous support from the resource people and individual mentors from different countries, have helped participants develop their own EE and ESD change project based on their institutional context. We hope that they will be able to implement their change projects and to make the real changes happen at their workplaces, so as to respond to the national strategy for sustainable development in the longer term.

Acknowledgements

Thanks to SIDA for generous support for this ITP and many thanks to the organisation team, all resource persons, colleagues and participants in Sweden, India, South Africa and China.

References

Ministry of Education. (2003). *National Environmental Education Implementation Guidelines for Students Grade 1–12*. Beijing: Beijing Normal University Publishing Group.

Qiaoling Wang. (2015). *Curriculum Review of China ESD in Formal Education* (ESD in Higher Education in Asia, pp. 43–52). Shanghai: Tongji University Press.

Shangri-la Institute for Sustainable Communities. (2014). Water school China: Project impact. Retrieved from http://waterschool.cn/projects/waterschool/impact/.

13 Sustainability from theory to practice

Chinese New Year as an avenue for sustainability education

Md Saidul Islam

Introduction

'Staying clean and green in Singapore remains as important as ever, because today we have a bigger population, we consume a lot more energy, we generate a lot more waste material' (Lee, 2014). This quote was taken from a speech by Singapore's prime minister, Lee Hsien Loong, at the launch of Clean and Green Singapore 2014 has highlighted the imperativeness of a sustainable environment to support both the current and future generations. Indeed, Singapore has come a long way since its first 'Green Movement' 50 years ago when Singapore's founding father Mr Lee Kuan Yew planted the first tree (*The Straits Times*, 2013). Recently, the government has set up a strategic environmental initiative – Singapore Sustainable Development Blueprint (SSDB) – that aims to map out a long-term plan to achieve the twin objectives of economic growth and good living conditions. The blueprint requires the three-way cooperation of the government, the private and the people sectors (Ministry of the Environment and Water Resources, Government of Singapore, 2014).

The SSDB is an example of a state-led Ecological Modernisation (discussed later in this chapter) in which the state takes initiatives to transform capitalism towards green practices through environmental policies, legislations, educations and enforcements. A central question arises: does this state-led Ecological Modernisation have any impact on the culture of consumption and vice-versa? To examine sustainability in Singapore from theory to practice, in line with this 'tripartite' framework and promotion of SSDB, we have conducted a survey among the students of HS0301: Environmental Sustainability course of Nanyang Technological University (NTU) around the Chinese New Year Celebration (CNYC) in Singapore. Through this study, we aim to see how celebrations like Chinese New Year can be an important avenue for sustainability education focusing on, for example, how the seemingly insignificant actions by individuals can have far reaching implications for our environment in the future. The study will shed light on how to teach sustainability in class (in theory) and reconnect students with nature and society (in practice).

The central argument of this chapter is that Ecological Modernisation is not limited to a top-down approach dominated by state administration or institutional reforms. Spaargaren and Cohen (2009) advocated that a sociologist needs to pay equal attention to the micro approach of understanding civil society based on consumption behaviour, lifestyle and daily routines as an important and complementary force to production in order to reconcile with environmental sustainability. Unlike the usual macro approach of Ecological Modernisation that has placed voluminous emphasis on institutional and policy change, our findings show that individuals also have great potential to exercise sustainable practices within their domestic sphere. We argue that bottom-up initiatives by individuals can also play a crucial role in making CNY greener. The active participation of individuals can transform CNY into a sustainable festival that benefits the masses. The starting point of this research is, therefore, to understand consumer agency in working towards a sustainable Chinese New Year in Singapore. In particular, the chapter investigates how NTU students experienced their CNY and what are the strategies they can employ to make CNY more sustainable and greener.

Following this introduction, the chapter first provides a background on Singapore's environment and culture highlighting the sustainability challenges and the celebration of Chinese New Year. Following this, Ecological Modernisation is examined as a framework. The methods of data collection on the Chinese New Year Celebration in Singapore and the key ideas are presented as three categories: consumptions, productions and institutions. The chapter highlights how Chinese New Year can be an important festive occasion for a wide range of environmental damages, while paradoxically offering an important avenue for sustainability education.

Background: Singapore and its environment

Singapore and its sustainability challenges

Singapore is a small nation state with a population of 5.3 million and a land area of 723 km (Department of Statistics, 2018). Known as one of the Four Asian Tigers, Singapore is considered as the wealthiest nation in the world with a capital GDP at market price of S$63,050.00 (Singapore Economic Development Board, 2012). Governed by the prudent People's Action Party (PAP) since its independence in 1965, the young nation has generally enjoyed a high standard of living and economic growth. A key contributor to Singapore's stable economic, physical and social infrastructure is its strategic geographical location which has cushioned Singapore from many, if not all, major natural disasters that have plagued its neighbouring countries for centuries. For example, Singapore was left virtually untouched by the 2004 Indian Ocean Earthquake and Tsunami which claimed more than 275,950 lives concentrated in South East Asian countries such as Indonesia, Sri Lanka, Thailand, Burma and Malaysia (World Vision, 2004). Historically, Singapore has been virtually spared from major natural and man-made catastrophes that have impacted on its geographical neighbours. However, Singapore is now being confronted with a number of environmental challenges that potentially increase its vulnerability. Three such challenges are haze, flash floods and climate change.

Haze

Entering the twenty-first century, Singapore has experienced a proliferated number of 'environmental problems' both directly and indirectly induced by humanity. Since the 1970s, the South East Asia Region has experienced the occurrence of haze every few years due to the burning of forests in Indonesia. However, between 2006 to 2013, haze from forest fires in Indonesia has ritually engulfed both Singapore and Malaysia yearly (Remember Singapore, 2013) at an unconventionally hazardous level. In June 2013, a state of emergency was declared in two districts of Johor Malaysia accompanied by the closing down of schools when the PSI level skyrocketed to 746 (Hussain & Teo, 2013). Singapore, on the other hand, has experienced an unprecedented record-breaking PSI level of 401 (Feng, 2013). Not only has the prolonged hazardous level of haze in Singapore threatened to delay the reopening of schools and sporting events (Chen, 2013), it has also severely damaged Singapore's economy such as the tourism sector, eateries, as well as Singapore's reputation as a business hub (Low, 2013). It was reported that the current haze is costing Singapore US$3 million a day before taking into account the inflation rate (Toh, 2013). Since Singapore's economy is heavily dependent on tourism and the service sector, the recurring prolonged hazardous haze has proven to be unaffordable by Singapore's delicate economy (Islam, Hui Pei, & Mangharam 2016).

Other consequences of the haze include haze-related health issues to which children and the elderly were most susceptible. Starting from June 2013 (June 2013 was the month when the haze problem reached its critical peak) to October 2013, more than 17,000 patients had claimed up to $50,000 in subsidies for haze-related ailments (Poon, 2013). The haze issues also resulted in diplomatic tensions between Indonesia and its neighbours when affected countries like Malaysia and Singapore pressed Indonesia to take action in resolving the illegal burning of forests (Saragi & Harahap, 2013).

Flash floods

Apart from air pollution, Singapore also saw an increased number of flash floods since 2010 that were largely caused by abnormally high rainfalls outside of the monsoon season compounded with blocked drainage systems (Singapore (AFP), 2010). The initial flash floods that first occurred in June 2010 have severely damaged businesses including the Central Business District area of Orchard Road. The initial flash floods have submerged underground car parks and caused traffic disruption. One of the food court chain operators, Kopitiam, who operated a chain in the worst hit building, Liat Tower (Liang, 2010), has reported a loss of $500,000 in infrastructure on top of damaged perishable food items (Liang, 2010). The rare occurrence of flash floods was attributed to climatic changes in the region. A climate expert interviewed by the local newspaper *The Straits Times* quoted that the abnormal rainfalls were caused by a weather phenomenon such as a typhoon in the region (ABC News, 2010). The increased frequency of flash floods in Singapore from 2010 to 2013 has prompted the government to implement extensive flood alleviation and prevention, flood protection and continuous improvement measures across the Singapore island (Public Utilities Board, 2011).

Climate change – dry spell

While Singapore has experienced unexpected heavy rain and flash floods in the past three years, the beginning of 2014 marked another extreme end of climate change. Since mid-January 2014, Singapore has encountered record low rainfall with February 2014 recorded as the driest month since the year 1869 (Channel News Asia, 2014a). During the month of February 2013, only seven days of short showers were recorded which was not sufficient to break the dry spell (Channel News Asia, 2014b). There was also a significant increase in the number of vegetation fires in the first seven weeks of 2014 in Singapore. It was reported at 99 cases as compared to 22 in the same period in 2013 (Seet, 2014). The continuous dry spell meant that warmer weather conditions (Channel News Asia 2014) would prevail forcing the government to step up efforts to issue advice to execute water saving practices (Seet, 2014). Of deeper concern to ordinary citizens was the threat of mosquito breeding and other health related issues (Lim, 2014). Without rain, the dry and warm weather would encourage mosquito breeding (Lim, 2014). A doctor has also reported seeing increased numbers of patients with respiratory illness and eczema. The prolonged dry spell was extended till the middle of March 2014.

Although the major environmental problems in Singapore have yet to claim massive numbers of deaths, the disruption to the Singapore economy, environment and social lives is immense. Singapore may not be directly responsible for some environmental problems such as the haze but the nature of environmental problems and their consequences do not respect nation state boundaries. Indeed, the various environmental problems that Singapore faced are not to be taken lightly. Furthermore, the highly industrialised nation state of Singapore was ranked ahead of other highly industrialised countries in twenty-seventh place among 137 countries for the most carbon emission per capita (OECD/International Energy Agency, 2011). For example, Japan is ranked in thirtieth place (OECD/International Energy Agency, 2011), Hong Kong at forty-second (OECD/International Energy Agency, 2011) and France at fifty-third placing (OECD/International Energy Agency, 2011). Therefore, we see a crucial need for Singapore as a 'victim' and a 'contributor' of environmental problems to tackle environmental issues both in the short and long run with a top-down governmental approach as well as bottom-up community-based civil engagement.

Historically, Singapore has been a patriarchal state where citizens rely heavily on the government to manage both macro and micro issues. The people of Singapore are overly pampered by the top-down administration, by the government such that civic movement has been weak. This is a possible reason for Singapore's sluggish Green Movement by bottom-up initiatives when juxtaposed against other industrialised nations in Europe (Wong, 2012). While the Singapore government has reacted relatively swiftly and efficiently to the above-mentioned environmental problems, the mere mitigation by the government alone is insufficient to ensure a stable physical and social environment in Singapore in the long run.

With the increased environmental awareness catalysed through education and the reality of climatic impacts hitting directly on Singapore's doorstep, it is timely for individuals in Singapore to take responsibility and action to ameliorate climate change and the environmental problems. In view of this, we have used sociological theory of Ecological Modernisation to analyse a case study of the Chinese New Year Celebration (CNYC) – a major festive celebration that has great impacts on both consumption and production of goods and services in Singapore.

Chinese New Year Celebrations and the Environment

The Chinese New Year 华人新年 (*hua ren xin nian*), often known as Lunar New Year or the Spring Festival is the most celebrated festival among the Chinese each year. Countries like Taiwan, China, Hong Kong and Singapore celebrate the festival through various public events and family gatherings. It is a time when the Chinese bid the past year goodbye and celebrate the coming of a new year, which normally lasts for half a month (Chang, 2010). During this period, Chinese hold on to various traditions and practices, while others are replaced and modified. One of these traditions is reunion dinners, normally on the eve of Chinese New Year. This is a time where families gather to have a meal together, normally held at the homes of the elderly. The wife of the family would have their reunion dinners with her husband's family in the first instance, then visit her own family.

Spring cleaning 大扫除 (*da sao chu*) is another tradition whereby Chinese would clean up their houses, throw away unwanted or old things and decorate their houses with 春联 *chun lian* (auspicious words written on red papers). Decoration may be complemented with plants that bring prosperity, such as the money plant. In the case of Singapore, makeshift stores especially in the heartland areas are set up, selling decorations and Red Packets. One would be overwhelmed by the amount of decorations sold in these shops, which are mostly in red. Chinese believe red is an auspicious and lucky colour, associated with the legend about the New Year monster 'Nian' which is scared of the colour red. As such, Chinese try to follow the tradition of wearing auspicious colours like red to bring in luck. The giving of Red Packets is also a symbolic representation of blessings and luck. This practice originated hundreds of years ago where parents would put some amount of money called 压岁钱 (*ya sui qian*) under their children's pillows on the eve of Chinese New Year. It was believed that the money would bring blessings and protection to their children in the coming year (Siu, 2001). Hence, this tradition is carried on today in the form of parents and senior family members giving out Red Packets to the younger family members as a form of blessing.

Aside from these traditions, many people are debating how social changes such as the normalisation of nuclear families have caused erosion of cultural practices during Chinese New Year. However, in the context of Singapore and China, many traditions and practices are still intact since Chinese New Year is religiously celebrated every year, be it at home, work or even school. Although it is increasingly more difficult for families to gather on the eve of Chinese New Year for reunion dinners, many of them chose to gather before the eve. Some believe that modification and readjustments in traditions do not actually mean that the values behind them are forsaken but instead represent 'a spirit of give-and-take' evolving among the Chinese (Tan, 1998).

Furthermore, it can be seen how the streets of Chinatown during the festive season were always filled with crowds of people either immersing themselves in the festive mood or purchasing New Year goodies. Chinese New Year goodies neatly placed in plastic containers, *bak kwa* (barbecued pork slices) and decorations including lights are the common things seen during Chinese New Year. In China's case, many Chinese nationals from all over the world would make an effort to fly back to their own country to celebrate the season, incurring large amounts of carbon emission through air travel. Travelling by air has been a big concern among many scholars because it has been increasing at a rate faster than the introduction of fuel-efficient air travel (Rosenthal, 2013). Moreover, the season has become largely commercialised, with wider media attention with the Spring Festival Gala telecasted on China Central Television (CCTV), trips to temples, tour packages, meal packages in restaurants and even films produced in relation to the festival (Ren, 2003). These programmes and ways of celebration generated large amounts of pollution and harm to the environment.

As discussed above, streets lined with vendors and lights, countless plastic containers and paper decorations, fireworks and wasted food, carbon emissions from air travel are just some of the pollution generated from the celebration of Chinese New Year incurred by people around the world. For instance, noise pollution is generated when media events are held; a large amount of electricity is needed for lights and decorations, while a large amount of carbon is emitted from air travel. Hence, CNY is paradoxically a time of joyous celebration and a time where the environmental problems escalate. In view of the paradox, this study analyses how Chinese New Year could be a site for environmental sustainability by employing the concept of Ecological Modernisation.

Theoretical framework: Ecological Modernisation

Ecological Modernisation is an environmental theory that emerged in the 1980s in Western Europe (Mol, Spaargaren, & Sonnenfeld, 2009). It has its roots in the history of environmental social science and has emerged as a policymaking perspective (McLaughlin, 2012). Ecological Modernisation was first launched 30 years ago at the Berlin state parliament for a study on 'Preventive Environmental Policy'. Built on the foundation of the 'greening of industry' theory by Joseph Huber, Ecological Modernisation has evolved into a major theoretical praxis for various environmental movements with economic, social, political and environmental implications (Buttel, 2000; Mol & Janicke, 2009). It is essentially a macro approach in looking at environmentally sustainable strategies with state involvement and practices at the organisational level.

Ecological Modernisation as praxis and an environmental discourse today is shaped by three main periods that have delineation purposes on the path international environmental policies should take. The first was the release of the 'Brundtland Report: Our Common Future 1987' (Mol, Spaargaren, & Sonnenfeld, 2009), the report for the Global North to reconcile the goals of both environmental concerns and economic development. It even prompted environmental non-governmental organisations (NGOs) to reframe their policies without sacrificing either goal. The fundamental aim of Ecological Modernisation is to adapt capitalism to achieve balance between economy, environment and social equity. The notion of bringing together the three institutions as 'complementary' (Mol & Janicke, 2009) instead of 'adversaries' (Mol & Janicke, 2009) under this report was introduced in the World Commission on Environment and Development (WCSD). Sustainable development was conceptualised here as a 'development that meets the needs of the present without compromising the ability of the next generation to meet their needs' (Estes, 1993).

As a way to practice sustainable development, there is a need for firms, either private or public, to review their production strategies to include environmentally sustainable practices. At the same time, Ecological

Modernisation emerged as a viable and valid solution to achieve sustainable development in the longer term. The second period was marked by the 'World Summit on Environment and Development 1992' where contemporary environmental problems such as climate change are brought to the attention of the global media and thus were widely reported. This set the stage for 'multilateral environmental agreements' (Mol et al., 2009, p. 5). Lastly, Al Gore's environmental activism on tackling climate change saw the extensive use of global mass media and advocating green consumerism and 'sustainable production' (Mol et al., 2009, p. 6). According to environmentalist Arthur P. J. Mol, Al Gore's climate change film *An Inconvenient Truth* (Gore, 2006) combines 'a strong doomsday storyline with an equally powerful invitation for individuals to demonstrate their moral concerns and translate these commitments into concrete actions of sustainable citizenship and consumption'.

The brief historical account of Ecological Modernisation as a practice of sustainable development reflects both a macro and micro approach to Ecological Modernisation in working towards a sustainable Earth. To further elaborate on the theoretical framework, there are two broad ideological mechanisms proposed by Ecological Modernisation. The first is free-market environmentalism, which does not question the existing economic and political structures but opts for 'ecologising the economy' (Mol et al., 2009). Examples include CERES (Coalition for Environmentally Responsible Economies) or corporate policies like Pollution Prevention Policy (3P) that advocates for environmentally-friendly production and waste-treatment processes. The free-market environmentalism argues that capitalism is flexible enough to embrace a movement towards achieving 'sustainable capitalism' (Buttel, 2000, p. 61) and the best mechanism to deal with environmental issues. The healthy competition between capitals alongside with stable political environment can help achieve ecologically efficient production and consumption processes with a focus on 'pollution prevention' (Buttel, 2000, p. 61).

Ecological Modernisation also argues for further modernisation also known as 'super-industrialization' (Buttel, 2000, p. 61). This brings us to the second mechanism of Ecological Modernisation called the reformist sustainable development. Reformist sustainable development seeks to integrate economic or state policies with sustainable development. There is a collaborative effort to reform policies by political actors locally, nationally and internationally in order to achieve the goals of economic growth, environmental sustainability and social equity. It not only involves a top-down approach but also emphasises on bottom-up initiatives like green consumerism. An example is the practice of 3Rs by consumers. To create a 'global civil society' (Mol et al., 2009), the civil society is strongly encouraged to participate actively in environmental decision-making processes. This is a key element in the transition from a centralised, bureaucratic state to one that integrates with the civil society.

The underlying principle behind Ecological Modernisation is 'ecological rationality' (Mol et al., 2009) where international or state policies are framed in consideration of the environment. In other words, behind any policy decision-making process, eco-efficiency and the limitations of environmental policies are rationally and thoroughly assessed. The ecological sphere is therefore of equal importance as 'political and economic spheres' (Mol et al., 2009, p. 7). This is termed by Mol as the 'emancipation of ecology' (Buttel, 2000, p. 61). Ecological Modernisation proponents believe that achieving economic development without abandoning environmental goals is possible. In fact, the concept argues that a healthy environment is a 'precondition' (Seippel, 2000, p. 288) for economic growth. Rather than viewing it as a clash of interests, both ecological and political economic spheres are essentially 'mutually reinforcing' (Seippel, 2000, p. 288). Proponents like Dryzek (1997) argue that Ecological Modernisation is a concept that requires capitalism to be restructured and guided 'along environmentally sound lines' without an entire overhaul of the system. To Dryzek, Ecological Modernisation enables economic goals to be achieved such as through green products. This theory works within the framework of modernity where progress can be made without compromising ecological concerns (Seippel, 2000, p. 291).

Ecological Modernisation is incomplete without emphasising the role of science and technology. One of the five core criteria in the Ecological Modernisation discourse is to explore how science and technology can be redefined to mitigate today's environmental issues (Mol & Sonnenfeld, 2000, p. 6). Although science and technology is being blamed for today's environmental issues, technology can also contribute greatly in mitigating the deterioration and prevention of environmental problems. The sphere of science and technology under Ecological Modernisation is often used as a strategic environmental management tool in industrial ecology. This includes cleaner, more efficient technologies and production practices (Hawken, 1993; Ayres, 1998) particularly applied to manufacturing industries. The policy reforms coupled with the production modification by the private industries help to internalise the environmental externalities (Buttel, 2000).

Ecological Modernisation advocates for two approaches. This approach of technological advancements to achieve ecological goals is what Hajer coined as 'techno-corporatist' (Mol & Janicke, 2009, p. 20). The second core criterion is the 'importance of market dynamics and economic agents' (Mol & Sonnenfeld, 2000, p. 6). Every agent involved in the production and consumption processes (financial corporations, producers and consumers) needs to participate actively and cooperatively with political institutions in framing environmental policies. The third criterion is the shift in governing style of nation states that entails states to eschew top-down administration of a 'controller state' (Bailey, Gouldson, & Newell, 2011). Instead, states act as 'facilitators' (Bailey et al., 2011). Greater autonomy should be given to citizens, non-governmental organisations and/or multinational corporations to negotiate and reform environmental policies, thus creating a 'state-society synergy' (Buttel, 2000, p. 63). The state should also 'stimulate social self-regulation' (Buttel, 2000, p. 61).

The fourth criterion is the change of social movements' 'position, role and ideology' (Mol & Sonnenfeld, 2000) whereby environmental movements align themselves with the global economic framework instead of the previous, more critical position towards economic growth. Environmental NGOs on the other hand are expected to proactively cooperate with the state, citizens, corporations or other institutions. The fifth and last criterion is the modification in how we approach environmentalism. Ecological Modernisation itself is a critique to 'radical environmentalism' (Buttel, 2000, p. 61). Framing economic and ecological interests in an opposing position is an unviable theorising position. The 'new environmental discourse' (Seippel, 2000, p. 291) is what Al Gore suggests as 'greening' (Mol et al., 2009, p. 6) within the bounds of economic development. Using Ecological Modernisation as our theoretical framework, this chapter employs the method of green discourse to engage the civil society to exercise some strategies for environmental sustainability. Our group believes that every individual has the potential to contribute to environmentalism. As such, our group decided to apply the Ecological Modernisation framework to our case study of a major festive celebration in Singapore-Chinese New Year Celebrations (CNYC).

Methods and procedures

Along with a brief review of sustainability challenges confronted by Singapore and the discourse of Ecological Modernisation to transform a society – albeit theoretically – towards a 'green' one, the study incorporated an empirical analysis of actual practices, to understand how students in Nanyang Technological University (NTU) engage in environmental sustainable practices during CNY. The samples were undergraduate students taking a NTU module on environmental sustainability. A total of 106 students were asked to respond to a general question through email with regards to the topic on environmental sustainability and CNY. The question was: *How can environmental sustainability be realised and reinforced through the Chinese New Year?* To provoke further thoughts and discussions, the respondents were instructed to pay attention to two things: (a) they can focus on how consumption is to be environmentally sound, socially responsible and culturally friendly; and (b) how CNY can be an avenue for sustainability education.

The study gathered the primary data from 92 email replies. Out of the 92 replies, we further categorised their responses into the following sections: consumption, production and institutional approach which are discussed in detail in the next section.

Findings and analysis

The research findings have been grouped into three major themes: consumption, production and institutions embedded in the context of Chinese New Year Celebration in Singapore.

Consumption

In Giddens's structuration theory, he argues that social behaviour and practices are not situated in a 'social vacuum' but are time and space specific (Beck, Giddens, & Lash, 1994). True to our findings, environmental problems and their possible improvements are intimately embedded in the cultural and physical space where individuals can reflexively exercise their agency within the structure to create a more sustainable Earth. Such agency is closely linked to everyday domestic lifestyles such as food, transport and housing utilities.

The majority of our respondents have focused on improving environmental sustainability from the perspective of domestic consumption. We summarise our findings on improving environmental sustainability with regards to consumption into four major categories: food, housing and utilities, Ang Baos (Red Packets) and transportation. Each type of consumption is elaborated in the context of the common practices and values embedded in the cultural traditions of celebrating Chinese New Year in Singapore.

Consumption of food

Reunion dinners and stock piles of Chinese goodies are annual traditions where immediate or extended family members gather to strengthen family ties. However, the reunion often entails environmentally unsustainable practices in the consumption of food which are symbolically attached to conventional Chinese values and traditions. Below are some major problems identified by our respondents with regards to environmentally unsustainable consumption of food practices. These problems are analysed in convergence to the traditions and values of Chinese New Year to better understand the cultural dimensions and our respondents also suggested solutions to these problems.

The first major environmentally unsustainable practice during Chinese New Year indicated by our respondents is the large amount of food wastage especially during reunion dinner. Consistent throughout all Chinese societies, reunion dinner is a pivotal ritual taking place on the eve of Chinese New Year where families gather to strengthen ties. Such reunion of kinship is accompanied by overly sumptuous food that is wasted at the end of the reunion dinner due to over-buying. This environmentally unsustainable practice is partially influenced by traditional

Chinese values practised during Chinese New Year. A common Chinese auspicious proverb that is used extensively to bless others during Chinese New Year is 年年有余 (*nian nian you yu*), meaning to be bountiful yearly such that there is more than sufficient. Furthermore, when a family hosts a reunion dinner for their extended families, food is often overly-prepared in order to 'save face'*.

Such symbolic values are translated into an over-estimation of the amount of food needed and result in large amounts of food wastage. To counter the problem of food wastage, many of our respondents suggested that families should make conscious budgeting and estimation of food required when shopping for food. Families should also have a change of mindset that the symbolic value of being bountiful should only stay at the symbolic level rather than be translated into practice. Others suggest that if there is unfinished food, it should not be thrown away but kept for consumption over the next few days or distributed to friends or the less fortunate as stated by one of the respondents, 'In my family, although we have leftovers, and we keep the leftovers and save it for the next meal or on other days. This helps to reduce the food wastage.'

The second undesirable practice with regards to food indicated by our respondents is the consumption of environmentally unsustainable food items. Such food items include exotic delicacies like shark fins, sea cucumber and salmon favoured by the Chinese. Exotic delicacies such as shark fins are often eaten during important events and gatherings such as wedding dinners and reunion dinners. Due to the high price that shark fins and the other exotic food items fetch, eating shark fins is more than often a symbol of wealth and luxury rather than for taste or nutrition. On the other hand, meal gathering during Chinese New Year often involves the tradition of 捞鱼生 (*lao yu sheng*) in which family members would use their chopsticks to scope up a dish made of mixed vegetables and raw salmon continuously while chanting auspicious words of blessings.

The consumption of exotic food items common during reunion dinners is therefore extremely disruptive to the ecological food chain and, paradoxically, Chinese New Year has become the peak season for animal cruelty. Most of our respondents have recognised this undesirable practice of consuming ecologically disruptive food and have made some suggestions to reduce this practice. The most straightforward suggestion which consumers can practise is to refrain from eating shark fins and other exotic food. This can be done through boycotting restaurants that serve shark fins or not ordering dishes that are ecologically disruptive, as commented by one of our respondents, 'boycotting shark's fin soup, turtle soup and other exotic dishes that are not environmentally sustainable'.

Based on market principles, should the demand for shark fins and other exotic food fall, producers will also supply less of these exotic foods that are often slaughtered inhumanly. In the case of 捞鱼生 (*lao yu sheng*), consumers can choose salmon that are from sustainable sources or substitute salmon with other seafood or meat as quoted below: 'Additionally, Yu Sheng foods could be more environmentally friendly by not using salmon, since this not only increases one's carbon footprint, but also reduces the consumption of an increasingly endangered species.' In fact, a few of our respondents also suggest reducing the consumption of meat and other products that have high carbon footprints. These include meat and luxury items like wine. Reducing the consumption of food with high food miles not only reduces the carbon footprint required to export and import them but it is also healthier to consume less of these foods and substitute them with more vegetables.

The last major problem concerning the consumption of food is the process of buying and distributing food items. In Singapore, plastic bags used for carrying groceries are given free to customers. During the Chinese New Year festival, the hike in shopping for groceries and other New Year goods simultaneously increases the use of free new plastic bags. Many plastic containers are also used for Chinese New Year goodies (snacks) packaging. Most plastic bags and plastic packaging, when thrown away or later used to contain disposed rubbish, are incinerated together with other waste. This produces harmful toxic gases and even when they are not burned, most plastic bags are not biodegradable. Similarly, plastic utensils are often used during reunion dinner and are disposed of after one-time-use to reduce the hassle of cleaning up. However, in exchange for convenience, the use of disposable plastic bags, packaging and utensils has resulted in much environmental degradation.

To reduce the problem with regards to the use of plastic bags, one respondent suggested that Singaporeans should cultivate the habit of using their own grocery bags instead of freely accepting plastic bags. As for disposable plastic utensils, many respondents proposed to substitute these with biodegradable utensils to reduce non-biodegradable waste. A better solution is refraining from using disposable utensils and opting for washable kitchen wares instead. Other respondents suggest that plastic containers can be reused to contain Chinese New Year goodies in the future or simply used to contain other food items.

Housing and utilities

大扫除 (*da sao chu*) or Spring Cleaning in English is a yearly tradition where families have a major clean up and tidying of the house. Chinese superstition claims that cleaning and sweeping the house during Chinese New Year is symbolically equivalent to sweeping wealth away, therefore Spring Cleaning is done before the first day of Chinese New Year. The purpose of Spring Cleaning is to maintain the cleanliness of the house for visitations by friends and relatives as well as to renew the look of the house. As the Chinese saying goes 旧的不去， 新的不来 (*jiu de bu qu xin de bu lai*) meaning the new does not come if the old does not go; Spring

Cleaning is also an opportunity to dispose of any unwanted items while welcoming new purchases. As such, a voluminous amount of old items and waste is discarded during Spring Cleaning. Some of these old items indicated by our respondents include old clothing, electronics and furniture that are still in good usable condition. Not only is it a pity to discard usable items, the festive season also witnesses a large amount of usable items that are not disposed of properly. For example, electronic items that can be environmentally hazardous are usually thrown together with other waste while discarded items are simply burned in incineration plants. Therefore, the disposal of old household items is not only a waste of money but an improper way of disposal which is environmentally hazardous.

To improve the process of discarding unwanted items during Spring Cleaning, our respondents provided several recommendations that are more environmentally sustainable and even socially equitable. Many of our respondents advocate that unwanted clothing should not be thrown away but recycled. This can be done through the selling of old clothing, usable furniture and even spoilt electronics to 'Garang Guni' (second-hand items collectors) which will then be sent for recycling. Alternatively, some respondents suggest donating usable clothing, electronics and furniture to charity organisations like the Salvation Army which will then be redistributed to the less fortunate communities in Singapore.

Items such as old clothing can also be reused by turning it into table cloths. For items that must be thrown away, a more environmentally sustainable practice is to separate the different items by material (e.g. glass, paper, metal, plastic) and recycle them in the neighbourhood recycling bins. In this way, not only can Singaporeans reduce waste during Spring Cleaning but also strengthen community ties in the festive season. Like any other festive season, many households would decorate their houses to lift the festival atmosphere. For Chinese New Year, a new lunar year is marked with an animal from the Chinese Zodiac consisting of 12 animals. As such, it is common to have household decoration based on the New Year's zodiac. However, this has caused excessive buying of New Year decorations as the zodiac animal changes annually. The Chinese New Year decorations are mostly made of paper and are sometimes discarded after the festive season is over.

To make Chinese New Year decorations more environmentally friendly, our respondents suggest that the decorations should be reused rather than buying new ones. If the zodiac animal changes, the decorations based on the zodiac animal should be kept for future use. Others suggest that families should simply decorate their homes with nonzodiac based decorations. Some of our respondents also suggest incorporating creativity in the decoration of houses by recycling scrap papers and unwanted household items to be made into decorations. In this way, consumers can reduce the money spent on household decorations and reduce the environmental effects for producing paper decorations and packaging as recommended by one of the respondents:

> Additionally, Chinese like to decorate their houses with festive accessories and buy Chinese New Year song albums to hype up the mood at their homes. It would be a better alternative to purchase festive decorations that can be reused for the following year and to buy virtual album through online medium such as iTunes store. These would greatly reduce the resources required and a healthier option for Mother Earth in the long run.

Lastly, a respondent suggested that households should switch to electric saving electronic appliances to save electricity. This includes using a fan rather than turning on the air-con when the weather is cooler. When not in use, electrical appliances and lights should be switched off. A few other respondents also share the consensus to make a conscious effort to save water during Spring Cleaning or the preparation of food as quoted below to:

> reduce the usage of water and electricity during this festive season by switching off the lights and water taps when not in use, switch on fan instead of air-con since it is quite windy these days and minimize the number of times that we open our refrigerator as it may consume more electricity.

As mentioned by one of our respondents, when individuals made small contributions to reduce energy usage and waste production, the combined effect will be significant in building a more sustainable lifestyle and a sustainable Earth.

Red Packets

A very important traditional practice looked forward to by children during Chinese New Year is the giving of 红包 (*Ang Bao*) or Red Packets. The Red Packets are usually given to young unmarried people such as children and students by parents and seniors. The Red Packets are given by putting money into a decorated red envelope made of paper. Many of our respondents recognise that the practice of giving and receiving money in new Red Packets yearly is environmentally unsustainable as these Red Packets are largely thrown away after retrieving the money inside as stated by one of the respondents; 'thousands and millions' of Ang Baos are given out each year, and the Red Packets, more often than not, are thrown away after the children or unmarried adults have collected the money. 'A radical change could be to give the money without the red packet to save paper.' However, such radical change is not culturally sensitive as it is considered rude for the recipient to expose the amount of money given when receiving the 'Ang Pao'. The problem of 'Saving Face' is also problematic should the amount of money be seen by the surrounding people.

A more acceptable and environmentally sustainable method suggested by some of our respondents is to reuse the Red Packets that are still in good shape. In this way, less paper will be thrown away while parents and seniors would not have to worry if they have enough Red Packets for future use. Once again, creativity can be incorporated in the spirit of recycling. A few respondents suggested people can 'DIY' (Do-It-Yourself) by using old red scrap paper to make their very own 'Ang Bao' instead of using commercially made ones. Others suggested that the 'Ang Bao' itself can be recycled by using it as part of the Chinese New Year Decoration or greeting cards and that we 'should not dump red packets once they are used, they can be recycled to make Chinese New Year decorations such as lanterns and cards'.

Transport

In the busy metropolitan Singapore, Chinese New Year is one of the few rare occasions where families and relatives reunite while friendship is strengthened. Known as 拜年 (*Bai nian*) or house visitations, this is an important traditional practice for people to visit their relatives and friends during the first few days of Chinese New Year to bless and catch up with each other. Because of Singapore's small geographical area, multiple visitations are made possible. However, making multiple trips to visit relatives and friends also means that there is high utilisation of transportation, often by cars. This increases carbon emissions as well as exhaust gases that pollute the air.

To reduce air pollution by vehicles, some respondents recommended that people should opt for the highly connected and extensive public transportation system in Singapore for house visitations. One respondent's experience said that public transport is also less crowded during the long public holiday of Chinese New Year as there is no rush hour. Choosing to take public transport instead of private cars helps reduce the carbon emission per capita during the festive season. Others suggested that if a large group of people are travelling together, the group can 'car-pool'; through coordination and maximising the number of people in each car as quoted below:

> Chinese New Year (CNY) usually entails large groups of people moving to and from places as there is the concept of CNY visiting where people gather at the houses of certain hosts. To keep in mind environmental sustainability, one could either car-pool or take public transport to minimize the number of vehicles on the road.

Taking public transport or using the method of car-pool can significantly reduce transport carbon emissions by people and at the same time reduce traffic congestion on the road, by making small changes to the way we use transport. Chinese New Year can be a festival that not only provides an opportunity for people to choose more environmentally sound transportation practices but at the same time strengthen social ties.

The above ways that our respondents suggest to improve environmentally sustainable consumption practices resonates with the Ecological Modernisation perspective of green consumerism. Green consumerism is a strategy of Ecological Modernisation that argues for the integration of ecologically sound products and services into the everyday complexity of lifestyle and household dynamics. The initiatives for environmental sustainable practices have to be realistic and within the grasp of individuals (Spaargaren & Cohen, 2009). Green consumerism comprises several dimensions that are applicable to our findings.

The first dimension is the practice of 3Rs – Reduce, Reuse, Recycle. The practice of 3Rs is previously tied largely to the emerging service and flow economy (Harper, 2008) where corporations reform their industrial practices to save costs and green market themselves. As shown in our findings, the 3Rs can also be practised by households making rational and do-able adjustment to their consumption lifestyle. The practice of reduction is largely applied to reducing the initial consumption of goods such as buying just enough food and saving on household utilities. Reduction is further practised on minimising the consumption of goods that have high carbon footprints such as luxurious imports of meat and alcohol. The entailing result is the reduction of waste such as food waste and carbon emissions. Also, the practice of reducing comes with many advantages such as increasing convenience in the cleaning up process after the consumption of (a necessary amount of) food, healthier diets and saving cost. The practice of reducing is therefore a highly practical incentive to encourage households towards an environmentally sustainable Chinese New Year.

The second 'R' referring to Reuse is applicable to highly structured and routine activities carried out during Chinese New Year such as decorations and giving out Ang Baos. Unlike the consumption of food where 'Taste' tends to be more flexible and actor dependent, the physical materials and material designs (Ang Baos, decorations and plastic containers for Chinese New Year goodies) used for the above routine traditions have not evolved much over the years. Furthermore, the materials involved in the highly structured activities during Chinese New Year are largely non-perishable goods such as paper and plastic containers. As such, the practice of Reuse is most appropriate for ritualistic activities involving non-perishable material goods with the added incentive of saving costs. On the other hand, the slight modification to reused items like making Ang Baos using red scrap paper is also a platform to showcase one's individuality and creativity.

From our findings, the last 'R' referring to Recycle is mostly associated with Spring Cleaning. During Chinese New Year, the act of recycling is probably the least significant in household consumption, revolving around selling or giving unwanted but usable items to recycling industries. As the act of recycling requires the further action of sorting out unwanted items into different material categories, a conscious effort and awareness is needed for consumers to make the initiatives.

The second dimension of green consumerism is coined as 'political consumerism' or 'citizen-consumer' which has its roots in green campaigns to transform production consumption chains (Micheletti, 2003; Oosterveer, 2007; Boström & Klintmann, 2008). 'Political consumerism' is a praxis where consumers are given a more active role to 'use their purchasing practices as a source of power for promoting sustainable transition'. (Spaargaren & Cohen, 2009, p. 263). This theory argues that the ethical values of consumers are a formidable force to drive Ecological Modernisation for both consumption and production chains (Spaargaren & Cohen, 2009).

The movement of political consumerism is associated with 'boycott' (Spaargaren & Cohen, 2010, p. 263) where consumers boycott goods that are unsustainable or unsustainably produced. A clear example from our findings is the 'boycott' of shark fins by either not buying shark fins or boycotting restaurants that serve them during the Chinese New Year season. Our respondents also encouraged the consumption of goods that are certified with sustainable sources and production processes. The green consumerism movement of political consumerism serves as an effective alternative to complex bureaucracies and expertise knowledge needed for policy making (Spaargaren & Cohen, 2010). At the same time, it has the potential to involve active participation by the masses. The third dimension of green consumerism identified from our findings is the maximisation of efficient utilisation of goods and services. This is in line with one of the characteristics of a sustainable society which 'work to become economically and environmentally efficient in all senses' (Harper, 2008). During Chinese New Year, the consumption of food and transport is most prominent with regards to the application of efficient utilisation. For example, buying food that can be kept over a few days maximises the consumption of food instead of generating much left-over food to be thrown away after a single meal. As for transport, a car-pool maximises the space and (the number of) trip utilisation of a single car. To further maximise the already available services, consumers can choose to substitute their transport by private cars with public transport to amplify space and energy efficiency per capita. Closely intertwined with the practice of Reducing, the maximum utilisation of goods and services is translated to the reduction of waste and carbon emissions with minimum interference to individuals' convenience.

Although the above analysis centred on how individuals' consumption is situated in a cultural context can be an important driver of Ecological

Modernisation, we would like to point out that consumption practices are intimately connected to production. In the following section, we will analyse how production also comes into play in promoting an environmentally sustainable society during Chinese New Year.

Production

Some of our respondents emphasise companies' ability to contribute to environmental sustainability through Corporate Social Responsibility (CSR). Corporate Social Responsibility (CSR) is known as the understanding by corporations/companies of the interdependence of 'business and society' and thus the importance of practices that promote 'public responsibility' (Blindheim & Langhelle, 2010, p. 109). Under the notion of 'public responsibility', companies and corporations can thus contribute to a sustainable environment by making responsible business decisions or advocating socially-friendly production and consumption practices. Environmental sustainability is a principle under CSR that can be adopted by companies while incorporating economic goals. Through CSR, committed companies can minimise pressing environmental problems such as resource wastage or non-biodegradable waste disposal arising from their 'business operations and interests' (ibid., p. 109).

Areas of focus suggested by respondents include minimising food and resource wastage; waste/plastic recycling and working with environmentally-friendly business partners. Retailers of Chinese New Year goods can source from suppliers who practice environmentally-friendly practices such as eco-friendly packaging. Restaurants can prevent ingredient wastage by designing reunion dinner menus that require chefs to make full use of any ingredients used. For example, chefs can use the red snapper's bones to boil soup instead of throwing them away after slicing the fish meat. Restaurants should also reinforce their policy of non-food wastage through fines. Major plastic ware manufacturers like SKP are encouraged to purchase the non-biodegradable plastic containers at low cost used to store Chinese New Year goodies. These plastic containers can be reused, recycled or sold as second-hand containers instead of being discarded.

Red packets or Ang Baos that are customarily given to employees and clients to promote the companies' goodwill can be made of recycled paper. Manufacturers can produce Ang Baos without the adhesive seal and the zodiac character of the year should not be printed so that consumers can be encouraged to reuse the Red Packets. All in all, corporations can help to spread environmentally sustainable messages by printing them on recycled paper and packaging them together with the Red Packets. If it is possible, companies/retailers should not distribute Red Packets. As stated by one, 'A better way would be to provide the red packets only when consumers request it instead of giving them out as complimentary free gifts.'

In convergence with the Ecological Modernisation paradigm, the above recommendations fit into the process of 'biomimicry' (Harper, 2008, p. 213). Companies or corporations alike play their part in reforming the capitalist economy by encouraging 'feedback loops' (Harper, 2008, p. 213) instead of continuing the practice of 'linear production and consumption' (Harper, 2008, p. 213). Moreover, by embarking on the process of 'biomimicry' (Harper, 2008, p. 213), we can materialise a 'service and flow economy' (Harper, 2008, p. 213) whereby products such as red packets and plastic containers can be 'remanufactured or recycled' (Harper, 2008, p. 213) by companies.

Another process is known as the 'radically increased resource productivity' (Harper, 2008, p. 213). Through the measures above, restaurants and manufacturers of Red Packets will drastically reduce the resources (ingredients; paper and ink) used for 'every unit of output' (Harper, 2008, p. 213) thus maximising profits while cutting production costs. For instance, Red Packet manufacturers can print the Red Packets on recycled paper. Likewise, the pressure exerted on red packet manufacturers by corporations which will purchase only Red Packets made of recycled paper exemplifies the notion of corporations fulfilling their role as 'social carriers of ecological reform' (Yee, Lo, & Tang, 2013, p. 103). An encompassing goal here is to achieve eco-efficiency under the directive of the 'economic logic' (Yee et al., 2013, p. 103). Therefore, the second core criterion characterising the Ecological Modernisation paradigm is fulfilled, the placing of high importance on 'economic agents' (Mol & Sonnenfeld, 2000, p. 6) involved in production and consumption processes to advocate environmental sustainability.

As argued by Yee et al. (2013), corporations who are successful in their endeavours in maximising productivity through cost-cutting environmentally-friendly practices serve as competitive role models for other firms in the same industry. This is particularly effective if companies choose 'industrial associations' (Yee et al., 2013, p. 116) in which the food and beverage or plastic manufacturing 'industrial associations' (Yee et al., 2013, p. 116) voice their opinions and successful experiences on environmental sustainable practices. Corporations can serve as 'information sources' (Yee et al., 2013, p. 116) for potential companies who may want to embark on environmentally-friendly business practices. The festive period of Chinese New Year therefore serves as a great opportunity for companies to initiate eco-efficiency practices. This parallels the Ecological Modernisation theory of 'ecologise the economy' (Mol et al., 2009, p. 7) through the free-market. The measures recommended by respondents implied the flexibility of capitalism to be reformed 'along environmentally sound lines' (Dryzek 1997) while attaining economic development.

Institutions

Various institutions can also step in to make CNY greener. This involves raising awareness through education programmes and media in individual schools and town councils initiated by the government. First, some students recommend that schools and neighbourhoods can generate awareness for environmental sustainability before, during and after CNY. This can be done via organising charity drives to encourage people to donate used clothes, old household appliances and books to the needy and less well-off in Singapore. Schools can also conduct educational campaigns with the help of grassroots welfare organisations as well as town councils which have the resources and social networks available for recycling programmes. Their tasks are to provide more recycling bins around the neighbourhood for the convenience of residents. Also, they can help promote environmental sustainability and green consumerism by carrying out educational campaigns about recycling, reducing and reusing to encourage environmentally-friendly habits.

Second, traditional and social media can promote environmental sustainability practices. Television and radio advertisements can reach out to a large population and are an effective tool to transmit knowledge and disseminate information to the public. One student shared his experiences with a video/advertisement created by Coca Cola named 'Create Happiness This Chinese New Year' which was uploaded on YouTube.

> It shows how a man's environmentally conservative act could be emulated. Towards the end of the video (from 24 seconds onwards), one male member of the family uses a decorated used Coca Cola box as a lion head to entertain his family members with the traditionally Chinese art form called the Lion Dance.

This example shows how participants in the CNY celebrations can be socially responsible and culturally friendly while also assuming the role of an environmentally sound consumer by reusing a box that would otherwise have been plainly tossed aside as waste. There are many more examples to show how media can raise awareness among the citizens and this is one such example. Through the help of popular socio-technical networks like YouTube, the popularity of CNY can be harnessed to promote the idea of environmental sustainability and used as an avenue for future sustainability education.

Third, Singapore has strong state involvement that regulates its environmental governance framework. Many students hope that the Singapore government can continue the relevant regulation and campaigns to make CNY greener. Some students encourage the government to be more proactive before and during the CNY season with initiatives such as 'to try to change their consumption behaviours by establishing relevant campaigns' and 'possibly provide monetary incentives' in the early stages so that the cultivation of such behaviours may become a 'nationalistic obsession for them' in the future. Some governmental grassroots welfare organisations such as Hong Kah North provided avenues for residents to recycle their used clothes or newspapers, glass bottles or metal cans and at the same time earn certain sums of money for their contributions. Also, most students agree that the Monetary Authority of Singapore (MAS) can continue to issue 'good-as-new' $2 notes which were 'clean and crisp notes which were retrieved immediately after the previous Lunar New Year'. This helps to save resources (paper/plastic) by not printing new notes, which is said to be able to power a hundred households for a month. One student actually hopes that the government can implement a recycling policy similar to ones found in The Netherlands

> where consumers pay 50 cents more when they purchase a 1.5 litre bottle of soft drink and if the consumers return the empty bottle at any of the automated machines beside the supermarket, the 50 cents will be returned to them as well.

Since the institutional approach tends to be state linked, there is a need for a more inclusive and participatory 'whole of society' approach to make Chinese New Year greener. The state thus functions as a 'facilitator' (Bailey et al., 2011) to ensure grassroots institutions promote environmentalism. This is also coined as the 'political modernization' (Mol & Sonnenfeld, 2000, p. 6) phenomenon. The Singapore government can initially assume a 'command-and-control' (Mol & Sonnenfeld, 2000, p. 6) role. However, this role is gradually outsourced to citizens, grassroots and non-governmental organisations in organising environmentally-friendly practices and approaches.

Political reforms such as the policy of monetary refunds to encourage recycling saw the merging of the economic goal to increase consumption with the promotion of environmentalism. Governance takes on a 'flexible and consensual style' (Mol & Sonnenfeld, 2000, p. 6). In this case, political regulation promotes the participation of producers and consumers alike in framing environmental policies. The festive period of Chinese New Year optimises the opportunity for the Singapore government to embark on long-term environmental policies that will grant non-political organisations the autonomy to participate actively in advocating and achieving environmental sustainability.

As mentioned earlier, one of the core criteria of Ecological Modernisation theory is to create a symbiotic relationship between the state and civil society. Moreover, the initial active role that the state has undertaken will send a message to 'economic agents' (Mol & Sonnenfeld, 2000, p. 6) that the state is willing to provide the 'political backing' (Mol, 2002, p. 105) for corporations and companies to align their business interests to environment sustainability interests. There is thus a gradual shift from top-down governing processes to 'partnership arrangements' (Berger, Flynn, Hines, & Johns, 2001, p. 59) with social institutions such as town councils, schools, welfare organisations and the media to bear greater responsibility and proactivity in encouraging environmentally-friendly practices. Therefore, the role of the Singapore government is to serve as a bridge to resolve the tensions between environmentalism and our economic agenda by taking on a 'social regulatory role' (Coffey & Marston, 2013, p. 183).

Discussion and conclusion

As discussed above, we have attempted to provide an in-depth analysis on the possible strategies to adopt during Chinese New Year using the framework of Ecological Modernisation. The contribution of our research is to provide an array of ideas on how the masses can play a part in environmental sustainability, instead of depending on the government. As analysed in the previous few paragraphs, our respondents have suggested ways in which people could contribute in sustaining the environment during the Chinese New Year festival. They have recognised how everyday lives could lead to negative impacts on the environment and thus highlighted the importance of education in raising awareness among Singaporeans. Suggestions include how campaigns and advertisements could be introduced into the public sphere, as well as family members educating each other on the importance of conserving the environment when they meet up during Chinese New Year. We found their suggestions on education largely feasible because they are bottom-up and most do not have to be reinforced by the state. Instead, these strategies encourage self-governance and community-based participation in sustaining the environment. This is in line with Foucault's concept of 'governmentality' (Foucault, 1991) where the state outsources the role of governing to the various institutions like family and schools.

In a way, people are given more agencies to govern themselves without direct administration from the state. As such, it is the purpose of this research to highlight the importance of education in influencing consumption patterns and everyday practices to adhere to environmental sustainability. Other than highlighting the importance of education, the research aims to contribute to the existing literatures on Ecological Modernisation. Our group noticed that most writings place immense emphasis on production in understanding Ecological Modernisation. However, our research chose to concentrate on the consumption aspect of Ecological Modernisation, where consumers can also contribute to the sustainability of the environment through their participation in the global market. Our findings show that bottom-up initiatives are equally feasible instead of having the corporations and the state to implement various policies and campaigns, especially during the Chinese New Year period. The efforts from both production and consumption are closely interlinked and should be encouraged to ensure environmental sustainability (Spaargaren & Bas 2000, p. 58). As such, we hoped that our research would have extended the implications of Ecological Modernisation from existing literature by focusing on the consumption aspect of Ecological Modernisation. Lastly, as Singapore is a multiracial society, our research can be extrapolated to other festivals celebrated in Singapore, such as Christmas, Hari Raya and Deepavali. We feel that various suggestions could be exemplified and used in other festivities, such as preventing food wasting common to all festivals. As such, the suggestion of reducing food wastage could possibly be used in all festivals to ensure the sustainability of the environment.

As mentioned earlier, Singapore is constantly under threat of environmental problems such as haze and climate change. These problems need to be addressed in a long-term hindsight. As such, we aim to raise awareness among NTU students and the public on the ways to make our environment greener using the case study of Chinese New Year Celebration. Although the celebration occurs only once annually, the group has explored the massive implications it has on both consumptions and productions of goods and services during the 15-day period which can directly and indirectly affect Singapore's environment.

Ecological Modernisation has proved to be a useful theory to be applied to CNYC as it relates to all individuals, organisations and governmental aspects of green and sustainable practices. Furthermore, our findings show that the bottom-up approach or civil engagement is as important as the top-down state or institutional approach in promoting environmental awareness in the production and consumption chain. As accurately pointed out by one of the student respondents, 'a man's trash could become another man's treasure', most respondents have agreed that the 3Rs (Recycle, Reuse and Reduce) are applicable in almost all situations during the CNY period. The act of recycling or donating food, unwanted clothes, cans and bottles to the garbage collectors or welfare organisations could also provide the poor with some income to live by.

Most importantly, education within households, in schools or neighbourhoods and using mainstream media can all go a long way down the road to protecting our environment. It takes time and commitment to achieve sustainable consciousness among social actors in society. However, from simple research like this, the findings are optimistic; we truly believe that our society can build a sustainable city if we set our time and effort to make our environment greener.

Note

* A colloquial slang used to mean avoiding humiliation.

Acknowledgements

For data collection, data analysis and during the writing phase, we gratefully acknowledge the crucial contributions of Dorcas Chang, Jertaime Wong Xi Ting and Lee Cheong Khi Wina.

References

ABC News. (2010). 'Flash floods stain Singapore's urban paradise reputation'. *ABC News*, 18 July. Retrieved 5 March 2014 from www.abc.net.au/news/2010-07-18/flash-floods-stainsingapores-urban-paradise/909584.

'At-a-glance: Countries hit'. (2005). *BBC News*. Retrieved 4 March 2014 from http://news.bbc.co.uk/2/hi/4126019.stm.

Ayres, R. (1998). *Eco-Restructuring*. New York: United Nations University Press.

Bailey, I., Gouldson, A., & Newell, P. (2011). Ecological modernisation and the governance of carbon: A critical analysis. *Antipode, 43*(3), 682–703.

Beck, U., Giddens, A. and Lash, S. 1994. *Reflexive Modernization: Politics, Tradition and Aesthetics in the Modem Social Order*. Cambridge: Polity Press.

Berger, G., Flynn, A., Hines, F., & Johns, R. (2001). Ecological modernization as a basis for environmental policy: Current environmental discourse and policy and the implications on environmental supply chain management. *Innovation, 14*(1), 55–72. Retrieved 11 April 2014 from www.tandfonline.com/doi/abs/10.1080/13511610125074?journalCode=ciej20#preview.

Blindheim, B.-T., & Langhelle, O. (2010). A reinterpretation of the principles of CSR: A pragmatic approach. *Corporate Social Responsibility and Environmental Management, 17*, 107–117. Retrieved 11 April 2014 from http://onlinelibrary.wiley.com/doi/10.1002/csr.235/abstract.

Boström, M., & Klintmann, M. (2008). *Eco-standards, Product Labelling and Green Consumerism*. London: Palgrave.

Buttel, F. H. (2000). Ecological modernization as a social theory. *Geoforum, 31*, 57–65. Retrieved 6 March 2014 from www.ic.ucsc.edu/~rlipsch/EE80S/Buttel.pdf.

Channel News Asia. (2014a). The driest month since 1869 for singapore. *Channel News Asia*, March 4. Retrieved 8 March 2014 from www.channelnewsasia.com/news/singapore/the-driest-monthsince/1020150.html.

Channel News Asia. (2014b). Dry spell leads to more. *Channel News Asia*, 25 February. Retrieved 8 March 2014 from www.channelnewsasia.com/news/singapore/dry-spell-leads-to-more/1012310.html.

Chang, M-Y. (2010). Lunar new year in Taiwan. *International Journal of Education through Art, 6*(1), 41–57.

Chen, F. (2013). Football, netball and sailing competitions postponed on health concerns. *The Straits Times*, 20 June. Retrieved from www.straitstimes.com/sport/haze-update-football-netball-and-sailing-competitions-postponed-on-health-concerns.

Coffey, B., & Marston, G. (2013). How neoliberalism and ecological modernization shaped environmental policy in Australia. *Journal of Environmental Policy & Planning, 15*(2), 179–199. Retrieved 11 April 2014 from http://dx.doi.org/10.1080/1523908X.2012.746868.

Department of Statistics. (2013). *Population Trends 2013*. Singapore: Department of Statistics, Ministry of Trade & Industry, Republic of Singapore.

Department of Statistics. (2018). Singapore in figures 2018. Retrieved from www.singstat.gov.sg/-/media/files/publications/reference/sif2018.pdf.

Dryzek, J. S. (1997). *The Politics of the Earth: Environmental Discourses*. Oxford: Oxford University Press.

Estes, R. J. (1993). 'Toward sustainable development: from theory to praxi'. *Social Development Issues, 15*(3), 1–29.

Feng, Z. (2013). Singapore has learnt 5 key lessons from haze crisis. *The Straits Times*, 7 July. Retrieved 11 April 2014 from www.straitstimes.com/the-big-story/the-haze-Singapore/story/singaporehas-learnt-5-key-lessons-haze-crisis-ng-eng-hen-201.

Foucault, M. (1991). Governmentality. In Burcell, G., Gordon, C., & Miller, P. (Eds.), *The Foucault Effect: Studies in Governmentality*. Chicago: University of Chicago Press.

Gore, A. (2006). *An Inconvenient Truth: The Planetary Emergency of Global Warming and What We Can Do About It*. Pennsylvania: Rodale.

Harper, C. L. (2008). *Environment and Society: Human Perspectives on Environmental Issues*. New Jersey: Pearson.

Hawken, P. (1993). *The Ecology of Commerce*. New York: HarperCollins.

'Haze – A burning issue for four decades'. (2013). *Remember Singapore*. Retrieved 4 March 2014 from http://remembersingapore.wordpress.com/2013/06/19/a-hazy-problem-for-40-years/.

Hussain, Z., & Cheng, W. T. (2013). Emergency declared in 2 Johor districts; Malaysia closes schools in several states, haze reaches south Thailand. *The Straits Times*, 24 June. Retrieved 5 March 2014 from www.straitstimes.com/st/print/1234767.

Islam, M. S., Yap, H. P., & Mangharam, S. (2016). Trans-boundary haze pollution in Southeast Asia: Sustainability through plural environmental governance. *Sustainability, 8*, 499; doi:10.3390/su8050499 (Switzerland).

Lee, H. L. (2014). Speech by Prime Minister Lee Hsien Loong at the launch of clean and green Singapore 2014. Prime Minister's Office Singapore. Retrieved 12 April 2014 from www.pmo.gov.sg/content/pmosite/mediacentre/speechesninterviews/primeminister/2013/October/speech-by-prime-minister-lee-hsien-loong-at-the-launch-of-clean-.html#.U0jeDPmSzU8.

Liang, A. (2010). Blocked drain caused Orchard Road flood. *My Paper*, 18 June. Retrieved 5 March 2014 from http://news.asiaone.com/News/AsiaOne+News/Singapore/Story/A1Story20100618-222731.html.

Lim, W. L. (2014). Dry spell leads to more vegetation fires, mosquitoes, health problems. *Channel News Asia*, 25 February. Retrieved 8 March 2014 from www.channelnewsasia.com/news/singapore/dry-spell-leads-to-more/1012310.html.

Low, A. (2013). Prolonged haze 'may hit Singapore's competitiveness'. *The Straits Times*, 24 June. Retrieved 5 March 2014 from www.straitstimes.com/the-big-story/asiareport/singapore/story/prolonged-haze-may-hit-singapores-competitiveness-20130624.

McLaughlin, P. (2012). Ecological modernization in evolutionary perspective. *Organization Environment, 25*, 178–196.

Micheletti, M. (2003). *Political Virtue and Shopping; Individuals, Consumerism and Collective Action*. London: Palgrave.

Ministry of the Environment and Water Resources, Government of Singapore. (2014). Sustainable Singapore: A lively and liveable city. Retrieved 12 April 2014 from http://app.mewr.gov.sg/web/Contents/ContentsSSS.aspx?Cont Id=1034.

Mol, A. P. J. (2002). Ecological modernization and the global economy. *Global Environmental Politics, 2*(2), 91–115. Retrieved 11 April 2014 from www.mitpressjournals.org/doi/abs/10.1162/15263800260047844#.U0etFE2KCUk.

Mol, A. P. J., & Janickle, M. (2009). Chapter 2: The Origin and Theoretical Foundations of Ecological Modernization Theory. In Mol, A. P. J., Sonnenfeld,

D. A., & Spaargaren, G. (Eds.), *The Ecological Modernisation Reader: Environmental Reform in Theory and Practice* (pp. 1–27). London: Routledge.

Mol, A. P. J., & Sonnenfeld, D. A. (2000). Ecological modernisation around the World: An introduction. *Environmental Politics, 9*(1), 1–14. Retrieved 6 March 2014 from http://dx.doi.org/10.1080/09644010008414510.

Mol, A. P., Spaargaren, G., & Sonnenfeld, D. A. (2009). Ecological modernisation: Three decades of policy, practice and theoretical reflection. In *The Ecological Modernisation Reader. Environmental Reform in Theory and Practice* (pp. 3–14). London and New York: Routledge.

'Ng Eng Hen says the authorities will continue to monitor situation, fine-tune contingency plans'. (n.d.). *The Sunday Times*, 7 July. Retrieved 5 March 2014 from http://news.asiaone.com/print/News/Haze/Story/A1Story20130708-435561.html.

OECD/International Energy Agency. (2011). CO2 emissions from fuel combustion-2011 highlights. National Climate Change Secretariat Prime Minister's Office Singapore. Retrieved 8 March 2014 from http://app.nccs.gov.sg/(X(1)S(zf04w245rqqseiuelebbz2b4))/page.aspx?pageid=158&secid=157&AspxAutoDetectCookieSupport=1.

Oosterveer, P. (2007). *Global Governance of Food Production and Consumption: Issues and Challenges.* Cheltenham: Edward Elgar.

Poon, C. H. (2013). No more subsidies for haze-related illnesses. *The Straits Times*, 1 November. Retrieved 5 March 2014 from http://yourhealth.asiaone.com/content/no-more-subsidies-hazerelated-illnesses.

Public Utilities Board. (2011). Overview of Singapore's drainage management approach July 2011. Public Utilities Board. Retrieved 11 April 2014 from www.pub.gov.sg/general/documents/overview_drainagemgmt.pdf.

Ren, L. (2003). *Imagining China in the Era of Global Consumerism and Local Consciousness: Media, Mobility, and the Spring Festival.* Electronic Thesis. Retrieved 7 March 2014 from https://etd.ohiolink.edu/.

Rosenthal, E. (2013). Your biggest carbon sin may be air travel. *New York Times*, 26 January. Retrieved 9 April 2014 from www.nytimes.com/2013/01/27/sunday-review/the-biggestcarbon-sin-air-travel.html?_r=0.

Saragi, B. T., & Harahap, R. (2013). RI dodges haze blame game. *The Jakarta Post*, 20 June. Retrieved 5 March 2014 from www.thejakartapost.com/news/2013/06/20/ri-dodges-hazeblame-game.html.

Seet, S. H. (2014). SCDF calls for public to be alert amid rise in vegetation fires. *Channel News Asia*, 2 March. Retrieved 8 March 2014 from www.channelnewsasia.com/news/singapore/scdf-calls-for-public-to/1017856.html.

Seippel, Ø. (2000). Ecological modernization as a theoretical device: Strengths and weaknesses. *Journal of Environmental Policy & Planning, 2*(4), 287–302. Retrieved 6 March 2014 from tp://dx.doi.org/10.1080/714038562.

Singapore (AFP). (2010). Singapore flood response not sufficient: Lee Kuan Yew. *Terra Daily*, 22 July. Retrieved 5 March 2014 from www.terradaily.com/reports/Singapore_flood_response_not_sufficient_Lee_Kuan_Yew_999.html.

Singapore Economic Development Board. (2012). *EDB Annual Report 2012/13.* Retrieved from www.edb.gov.sg/content/dam/edbsite/news-and-resources/resources/corporate-publications/Annual-Report-2012-2013.pdf.

Siu, K. W. M. (2001). Red Packet: A traditional object in the modern world. *Journal of Popular Culture, 35*(3), 103–125.

Spaargaren, G., & Bas, V. V. (2000). Lifestyles, consumption and the environment: The ecological modernisation of domestic consumption. *Environmental Politics*, 9(1), 50–76. Retrieved 11 April 2014 from http://dx.doi.org/10.1080/09644010008414512.

Spaargaren, G., & Cohen, M. J. (2009). Greening Lifecycles and Lifestyles: Socio-techological Innovations in Consumption and Production as Core Concerns of Ecological Modernisation Theory. In Mol, A. P. J., Sonnenfeld, D. A., & Spaargaren, G. (Eds.), *The Ecological Modernisation Reader: Environmental Reform in Theory and Practice* (pp. 257–274). London: Routledge.

Tan, S. S. (1998). Traditions can be changed without loss of values. *The Straits Times*, 26 January. Retrieved 11 April 2011 from www.lexisnexis.com.ezlibproxy1.ntu.edu.sg/ap/academic/.

The Straits Times. (2013). To mark 50 years of Singapore's green movement: Grow a plant out of a stamp. *The Straits Times*. Retrieved 12 April 2014 from www.straitstimes.com/breakingnews/singapore/story/mark-50-years-singapores-green-movement-grow-plant-out-stamp-20130713.

Toh, Y. C. (2013). Three clear lessons from hazy days. *The Straits Times*, 24 June. Retrieved 5 March 2014 from www.singapolitics.sg/views/three-clear-lessons-hazy-days.

Wong, C. M. L. (2012). The developmental state in ecological modernization and the politics of environmental framings: The case of Singapore and implications for East Asia. *Nature and Culture*, 7(1), 95–119.

World Vision. (2004). 2004 Indian Ocean tsunami. Retrieved from www.worldvision.org/disaster-relief-news-stories/2004-indian-ocean-tsunami-facts.

Yee, W-H., Lo, C. W-H., & Tang, S-Y. (2013). Assessing ecological modernization in China: Stakeholder demands and corporate environmental management practices in Guangdong province. *The China Quarterly*, *213*, 101–129. Retrieved 11 April 2014 from http://journals.cambridge.org/abstract_S0305741012001543.

14 Education for Sustainability – where do you go from here?

Chew-Hung Chang, Gillian Kidman and Andy Wi

Introduction

There have been a range of environmental projects and policies described in the chapters of this book, but there is no 'one-size-fits-all' solution as different countries have different social, cultural and environmental contexts (Chang & Wi, 2018). A successful action or strategy is not easily transferable across contexts. For instance, island states might favour resilience and adaptation to climate change over a focus on mitigation. Furthermore, we need to ask questions that connect the space, people and culture of a relevant issue at local, national and international levels (Adger, Arnell, & Tompkins, 2005; Massey, 2007).

In other words, we are interested in understanding EfS for whom, and EfS by whom, in addition to what an EfS curriculum looks like for these contexts. While this edited book volume has covered these various aspects, the chapters have also provided examples of how theory can be translated into practice and how some countries' schools and Higher Education Institutes (HEIs) are carrying out EfS.

Although the ontological differences in themes and contexts across the examples cited are apparent, there are some epistemological confluences to these discourses, as represented by the chapters. Indeed, the chapters almost always involved perspectives from multiple stakeholders, discourses on the economy, society and the environment and a notion of EfS for the future. The convergence of these ideas is aligned with the Brundtland Report's definition of sustainable development as 'development that meets the needs of the present without compromising the ability of future generations to meet their own needs' (Brundtland, 1987). By extension, EfS must equip learners 'with the knowledge and ways of thinking that' meet this aspiration about sustainable development (Cloud, 2004, 2010).

The United Nations Conference on Environment and Development held in Rio de Janeiro in 1992 recognised the critical role that education can play in sustainable development. But there is often an assumption that teaching about sustainability described within a well-designed curriculum is a necessary and sufficient condition to help our children engage sustainability issues in the future. Just talking about something does not mean that the idea has been communicated effectively. Teaching as an act does not assure that students have learned. Considering what and how our students should learn will offer us a more viable plan to attain the goals of EfS. 'Learning: The Treasure Within', commonly referred to as the 'Delors Report' of 1996 to the United Nations Educational, Scientific and Cultural Organization (UNESCO), at the International Commission on Education for the 21st Century, was premised on a holistic and integrated vision of education based on the four pillars of learning including:

1. learning to know – a broad general knowledge but also depth in a few subjects;
2. learning to do – to acquire not only occupational skills but also the competence to deal with many situations;
3. learning to be – to develop one's personality and to be able to act with growing autonomy, judgement and personal responsibility;
4. learning to live together – by developing an understanding of other people and an appreciation of interdependence.

In 'Climate Change Begins at Home', Reay (2005) took the critical view that the person on the street has as much a role to play in mitigating climate change impacts as governments and industries. He demonstrated how small changes in modern day living can culminate in a substantial reduction in greenhouse gases (GHGs) and hence slow down or halt human induced climate change. Some suggested changes to lifestyle include changes to travelling habits, reduction in the purchase of imported food and changing the light bulb to an energy efficient model. In his book, Reay suggested ways to assess the situation before making changes to lifestyle. In one example, he discussed the various options to reducing car use, including alternatives like hybrid cars and solar powered cars, all in terms of greenhouse emissions savings in comparison to a 4-litre petrol-powered car. Although individuals need to do something about climate change, it is not through taking big radical steps, but taking on gradual and sustainable behaviour; such as changing one's driving habits. The question then is, 'How can we encourage individuals to take on gradual and sustainable behaviour?'

While the impact of any individual's personal behaviour makes a significant impact on the environment (Stern, 2000), research has shown that increased awareness does not necessarily lead to action (Chang, 2014; Collins, Thomas, Willis, & Wilsdon, 2003). Perhaps changing individuals' mindsets by educating and encouraging them to play their part in mitigating climate change is the first step for individuals to participate in activities to mitigate climate change. Therefore, it is important to revisit some of the key learning points from the chapters and how they contribute to the three themes of knowing, doing and being in EfS.

Knowing

There is an imperative for the school curriculum to play a vital role in educating children to respond to pressing global problems, such as environmental degradation, climate change, economic challenges and sustainable development (Chang & Wi, 2018). Knowledge about these issues is important for developing a new generation of critical thinkers. There should, however, not be an overdependence on developing an ideal curriculum that provides holistic and comprehensive knowledge about the issues (Kagawa & Selby, 2012), but also a need to equip teachers with pedagogical readiness and awareness on the complexities involved in the teaching and learning about EfS.

One critical theme of sustainability education deals with disaster risk reduction (DRR). Hazards and disasters have been known to devastate the ecologies and natural habitats of living things as well as disrupting local economies, rendering already vulnerable communities to a state of physical and financial insecurity. As such, DRR seeks to emphasise the importance of raising awareness and knowledge about the capacities for individuals and communities to adapt to present and projected calamities that could threaten the sustainability of development. However, educators have raised issues concerning the lack of coverage in the curriculum of specific subjects and issues-based discussions and a lack of efforts in addressing the knowledge-behaviour gap. Acknowledging these issues in Chapter 6, the author argues from a postcolonial perspective that as hazards and disasters are largely influenced by their context, DRR researchers and educators should be cognizant of the fundamental limitations in their respective disciplinary worldviews and be critical of mainstream epistemologies that create artificial knowledge compartments. At the same time, an increased emphasis should be placed on the specificities of a place, its geography and its people when constructing knowledge in DRR. The author acknowledges that the official knowledge developed in schools has been largely influenced by the dominant techno-scientific epistemology which has hindered the development of plural knowledge and the contribution of equally relevant epistemologies. It is important to develop an educational experience that empowers the learner, the teacher and their communities in adapting to hazards and disasters.

The forces of globalisation have also shaped some countries to undertake extensive educational reforms in regards to EfS. Vietnam, as elaborated in Chapter 11, is experiencing challenges concerning the mismatch between the skills taught in tertiary institutions and the skills required in the job market. This has inadvertently resulted in a large percentage of university graduates being unable to secure jobs in their area of specialisation. In turn, the Vietnamese Ministry of Education and Training (MOET) has adopted three main approaches to modernising their education system, namely: (1) outward-looking education; (2) socialisation of education; and (3) job-hunting first approach. In particular, the job-hunting first approach seeks to address the existing mismatches in higher education in Vietnam that has caused high rates of unemployment in the country. In this approach, the Vietnamese policymakers and educators have initiated career orientations in high school activities to assist students in finalising their choices on specialisation that they will pursue at higher education institutions. At the same time, undergraduate students are encouraged to participate in practice-based modules so as to be acquainted to real work when entering the workplace. This rapid shift away from a liberal-arts education system towards a lucrative career-oriented education works to ensure that students' qualities are aligned neatly with the demands of the economy.

Infante and Jin (in Chapter 9) also stressed the importance of allowing learners to leverage on the practice with which they are familiar so as to foster a mindset of implementing sustainability in the decisions they make in their future careers. They explained that while EfS has been conceptualised and applied in different fields, it has not been explored thoroughly in power system markets taught in higher education (see Chapter 9) and they suggested that the concept of a process-based approach with the pillars of sustainability could be used to introduce Education for Sustainability in electrical engineering.

However, just because someone has knowledge does not mean their personal values or view of climate change can make them take action (Collins et al., 2003). Studies in the literature have shown that increased awareness does not lead to action (Collins et al., 2003), and having awareness about climate change is in itself not enough (Chang, 2014). Therefore, knowledge is not a necessary factor but an important pre-condition of EfS and ESD.

Doing

There is, however, a body of work that relates environmental awareness to students' attitude and behaviour (De Young et al., 1993; Hornik, Cherian, Madansky, & Narayana, 1995; Oskamp, 1995; Pelletier, Tuson, Green-Demers, Noels, & Beaton, 1998). Unfortunately, this empirical correlation does not explain how a large number of people remain indifferent to environmental action (Abrahamse, Steg, Vlek, & Rothengatter, 2007; Chang, 2008; De Young, 1988; Forester, 1988). This lack of consensus contributes to a continual search for ways to overcome the inconsistency of an individual's intentions and the resultant actions (Boyce & Geller, 2001; Darnton, 2004; Norgaard, 2009). So how do we get people to take the first step? How do we encourage people to even want to do something about the environment?

In Chapter 7, Yembuu and Getsel underscore the importance of fostering climate change awareness and understanding it from a young age as being the best strategy of altering behaviours and attitudes. Reflecting on the case of Mongolia, their analysis of the school curriculum reveals that a limited exposure to climate-related issues at the primary, secondary and tertiary levels has led to a lack of basic understanding and skills relevant to climate change adaptation. In the Mongolian system, climate-related issues are typically taught under geography and the natural sciences, therefore students who do not take these subjects/courses will have a limited knowledge of the importance of climate-related issues. Yembuu and Getsel also highlight that teacher knowledge and skills have been limited in teaching climate change adaptation, and cross-curricular activities are very limited in the national curriculum of education. They also point out that teachers' methodology is theoretically-oriented and recommend that teachers need more teaching aids to help them efficiently transfer their knowledge to students. Moving forward, they argue that climate change education should be taught more extensively across all levels and should be inclusive rather than discriminatory. An increased level of awareness, in turn, could lead to more environmentally conscious initiatives by students. Beyond just an argument for effective pedagogies, this chapter highlights the need for EfS to be well resourced and well designed. The same argument as highlighted in Chapter 6 on DRR can be applied here. In the absence of information, whether it is as complex as providing plural knowledges about DRR or having sufficient resources or a well-designed curriculum to teach about environmental issues, students are denied the agency to learn and subsequently do anything about what they have learned. Of course, an argument can be made that having sufficient resources is only a necessary but not sufficient condition. Echoing this view, Casinader & Kidman (in Chapter 8) recommend that curriculum needs to go beyond simply the teaching of facts but also on how to then use their knowledge to prepare for a sustainable future. They suggest fieldwork inquiry because it encourages participatory engagement in sustainability education by giving students the opportunity to go outdoors and safely explore their surroundings. To do so, they advocate for the employment of a *sustainability Education Fieldwork* that would stimulate the development of: (1) inquiry-based skills that foster a habit of being mindful of ones' surroundings; (2) affective skills of exploring the world by doing, and (3) personal skills and a sense of responsibility.

Fieldwork is a form of place-responsive pedagogy as it involves explicit efforts to teach in an environment of an environment, with the aim of understanding and improving that environment and its human-environment relations (Mannion, Fenwick & Lynch, 2013). How field inquiry can lead to action can be grounded in experimental and cooperative learning, involving the clarification of student values and a critical reflection that would enable the envisaging of a sustainable future. During the fieldwork inquiry, the teacher needs to facilitate the students' movement towards this goal. Subsequently, students should be encouraged to go beyond producing a written report, to take practical action to improve the situation. Indeed it is important for the student to act upon the short or long-term implications of the inquiry. This will help the learners develop inquiry-based skills, and be mindful of one's surroundings. Consequently, we also help students develop affective skills of exploring the world by doing, and cultivate a sense of responsibility.

Likewise, Likitswat (in Chapter 4) shared their experience on using built environmental modules to incorporate sustainable learning as well as active-learning for big classes at the tertiary level. These programmes were run by the teaching team including faculty members and teaching assistants, mostly graduated architecture students.

The ability to reflect is considered an essential element of EfS. Despite this, little is known about how reflective thinking can be identified, influenced or encouraged in the classroom. Chapter 10 described a study on how educators in a high school and in a HEI diagnosed their students' reflective thinking performance and facilitated reflection about sustainability. The chapter suggests that for reflection to work, it is important to establish links to the student's everyday life whenever possible, in order to increase reflective thinking performance. It is also imperative to confront students with the complexity inherent in many issues of sustainability and challenge them to deal with uncertainty.

Besides students, Wi (in Chapter 5) argues that the government should also be involved in EfS and ESD as climate change education policy is an important factor in garnering people to do something for the environment. His Grassroots Approach (GRA) suggests that there should be interaction between the top-down (government) and bottom-up (people), in terms of policy planning, information and implementation, public consultations, involvement, feedback and advisory. For the GRA to be effective, the author argues that it is essential that the government provides information during the implementation stage. Following which, the government should consult its citizens for an evaluation of their policy implementation. The author advocates that perhaps to get people to take action, the government (policy) should lead by example and have more communication and support for the people. Through the feedback stage, the information is then channelled to the advisory whose role is to assess the feedback and reformulate state policy accordingly. Although hypothetical, the author acknowledges that in order for this workflow to be effective, the government has to provide adequate direction and support systems through Grassroots Organizations (GROs) to incentivise people to adopt pro-environmental behaviour. People are motivated to learn new forms of knowledge only when they know that they can apply what they have learned to improve their local community (Schunk, Meece, & Pintrich, 2012). Therefore, it is important to provide meaningful learning to learners.

Being

Most governments tend to provide as much information about climate change as possible with the hope that in doing so, people might adopt sustainable behaviour. While most climate change education policy has been successful in creating awareness, it does not encourage people to behave sustainably. The reason is that people need to make sense and construct meaning from the information given before they can be empowered to take action.

Through a survey of students from the Nanyang Technological University (NTU) on Chinese New Year celebrations in Singapore, Islam (in Chapter 13) argues that the conception of state-led Ecological Modernisation in which the state adopts initiatives to induce capitalism towards green practices through policies and legislation should not be dominated by state institutions but should also account for bottom-up approaches. Specifically, Islam notes that individuals have great potential to exercise sustainable practices within their daily lives thus highlighting the importance of education in raising awareness among Singaporeans. His findings suggest that rather than having corporations and the state implement various policies and campaigns, bottom-up initiatives through family members and friends can influence an individual's consumption patterns and everyday practices to adhere to environmental sustainability. By contrast, Brendel (in Chapter 10) suggests confronting students with the complexity inherent in many issues of sustainability and challenge them to deal with uncertainty.

While research and reports have shown that environmental campaigns/events are effective in creating awareness and encouraging participants to take action, there is no information on whether the actions and behaviours are sustained or repeated. The literature on psychological studies shows that most people react to situations they consider personally relevant or when it affects their own livelihood (Moser & Dilling, 2004).

In addition, there is no 'magic bullet' in EfS and the effectiveness of education differs across countries. Research has shown that bottom-up or community-based approaches in sustaining the environment are important aspects in promoting EfS and sustainable behaviour (in Chapters 5 and 12). However, climate change mitigation remains mostly confined to education and awareness building, because many economic and especially regulatory instruments do not work effectively without enforcement and compliance.

Wi (in Chapter 5) notes that for nationwide climate change initiatives to succeed, it is critical to understand the public's perspective, social processes and human-environment interactions (Jordan, Hungerford, & Tomera, 1986). Therefore, he argues that a good EfS approach requires mutual interaction between the government and the people. Islam (in Chapter 13) also highlights that the bottom-up approach or civil engagement is as important as the top-down state or institutional approach in promoting environmental awareness in the production and consumption chain. Having this interaction will bring information from the government to the people and at the same time inform policy formulation at the national level.

Beyond state-society relations, collaborative programmes can also transcend national boundaries linking multiple institutions and organisations across the world for the exchange of efficient practices, policies, systems and knowledge. In Chapter 12, Qi highlights the International Training Program (ITP) in China which shows the cooperation between not only the government and its people but also experts from other countries. Supported by the Swedish International Development Agency (SIDA), and in partnership with NIRAS (a Swedish consultancy company), Environmental Education Centre in East China Normal University (EEC ECNU), Southern African Development Community Regional EE Program (SADC-REEP) and several other organisations in Sweden and China, the ITP has involved more than 180 environmental education practitioners and their organisations in Chinese formal education at different levels, including educational decision makers, curriculum developers, teaching researchers, school principals and teachers. By the end of the programme, each participant would finalise his or her EE and ESD CHANGE project based on the institutional context. Participants were expected to implement their projects in their workplaces to effect real changes. In addition, participants also stressed three important learning values which are *learning from doing, theory in practice* and *assessment as learning*. The result was a decentralised, outcomes-based curriculum which gave more responsibility and freedom to local educational authorities, individual schools and subject teachers. This provided teachers and educators with opportunities to develop their own curricula and work towards the expected outcomes. Nonetheless, the participants also expressed a few shortcomings in the programme. Some of these challenges include: (i) managing change; (ii) application of the lessons learned from the course to the local Chinese context; and (iii) EE and ESD evaluation and assessment in formal education. Moreover, the language barrier also makes it difficult for Chinese EE and ESD practitioners to communicate with the international community. The author notes that although these challenges were attended to by the organisers, the response of the organisers has been limited.

Conclusion

UNESCO defines curriculum as 'a systematic and intended packaging of competencies (i.e. knowledge, skills and attitudes that are underpinned by values) that learners should acquire through organised learning experiences both in formal and non-formal settings' (UNESCO, 2016). Moving forward, it is undeniable that a school's curriculum will be an integral tool in empowering the next generation of experts and leaders in devising innovative solutions to the world's most pressing problems, such as climate change and sustainable development. In a rapidly evolving world steeped in uncertainty and complexity, curriculum planning should not only be about each of the constituent outcomes of knowledge, skills and attitudes, but should also be a synergy of these outcomes that will enable the child to discern 'tested beliefs from mere assertions, guesses, and opinions' (Dewey, 1997).

Pedagogical and technological innovations at the global scale will certainly draw the boundaries on what can be done with the learning of EfS. However, there are developments beyond the sphere of formal teaching and learning spaces that we need to consider. The world is facing a range of global issues such as uneven access to education, social issues as a result of economic disparity across regions, and unprecedented environmental changes such as climate change. These issues will surely impact the learning environment of the rapidly evolving world. More importantly these are issues that are of relevance to the child that we are educating.

The 'quality of an education system cannot exceed the quality of its teachers' (Barber & Mourshed, 2007). Indeed, Casinader and Kidman (see Chapter 8) suggest that curriculum planning needs to go beyond simply the teaching of facts and reflecting the priorities of today's society. We now need to teach children to think critically about the nature of knowledge and to then use their knowledge to prepare for a sustainable future. Moreover, a discussion on the development of sustainability curricula would be incomplete without considering the changing roles of the important key stakeholders parents, teachers, curriculum planners and policymakers, as well. Teachers play a critical role in guiding students to think about the information that they come into contact with. As educators, we should first have a good understanding of the subject matter before we can even influence the learners' attitude and encourage them to take action (Chang & Wi, 2018).

This book has built on the previous book *Education and Sustainability: Paradigms, Policies and Practices in Asia* by the EfS Asia community. We have examined what EfS is, its applicability and its implementation around the world. The chapters have also discussed what an EfS curriculum looks like and we have given examples and strategies on developing a sustainability curriculum. Examples of how countries translate 'sustainability strategies' from theory to practice at the local, national and global scale were also described. In all, we have argued that the schema for understanding what EfS is all about does not detract from the four pillars of the UNESCO Delors report – learning to know, learning to do, learning to be and learning to live together. We seek to advance a deeper understanding of issues in teaching and learning in both education and sustainability, with a view to empowering our learners of tomorrow with the knowledge, skills, attitudes and agency to engage the issues in sustainability that they will face.

References

Abrahamse, W., Steg, L., Vlek, C., & Rothengatter, T. (2007). The effect of tailored information, goal setting, and tailored feedback on household energy use, energy-related behaviors, and behavioral antecedents. *Journal of Environmental Psychology, 27*(4), 265–276.

Adger, W. N., Arnell, N. W., & Tompkins, E. L. (2005). Successful adaptation to climate change across scales. *Global Environmental Change, 15*(2), 77–86.

Barber, M., & Mourshed, M. (2007). *How the World's Best-performing Schools Systems Come out on Top*. London: McKinsey & Company.

Boyce, T. E., & Geller E. S. (2001). Encouraging college students to support pro-environment behaviour: Effects of direct versus indirect rewards. *Environment and Behaviour, 33*(1), 107–125.

Chang, C. H. (2008). *Climate and Climate Change: A Singapore Perspective*. Singapore: McGraw Hill.

Chang, C. H. (2014). *Climate Change Education: Knowing, Doing and Being*: Abingdon: Routledge

Chang, C. H., & Wi, A. (2018). Why the World Needs Geography Knowledge in Global Understanding: An Evaluation from a Climate Change Perspective. In Demirci, A., González, R. M., & Bednarz, S. W. (Eds.), *Geography Education for Global Understanding* (pp. 29–42). Cham: Springer.

Collins, J., Thomas, G., Willis, R., & Wilsdon, J. (2003). Carrot, sticks and sermons: Influencing public behaviour for environmental goals. A Demo/Green Alliance Report for Defra. Retrieved 5 January 2014 from www.demos.co.uk/files/CarrotsSticksSermons.pdf.

Darnton, A. (2004). Driving public behaviour for sustainable lifestyles report 2 of desk research commissioned by COI on behalf of Department of the Environment, Food and Rural Affairs (DEFRA). Sustainable Development Unit of the Department for Environmental Food and Rural Affairs. The Government of United Kingdom.

De Young, R. (1988). Exploring the difference between recyclers and nonrecyclers: The role of information. *Journal of Environmental Systems, 18*(4), 341–351.

De Young, R., Duncan, A., Frank, J., Gill, N., Rothman, S., Shenot, J., Shotkin, A., & Zweizig, M. (1993). Promoting source reduction behaviour: The role of motivational information. *Environment and Behaviour, 25*(1), 70–85.

Dewey, J. (1997). *How We Think*. London: Courier Corporation.

Forester, W. S. (1988). Solid waste: There's a lot more coming. *Environmental Protection Agency Journal, 14*, 11–12.

Hornik, J., Cherian, J., Madansky, M., & Narayana, C. (1995). Determinants of recycling behaviour: A synthesis of research results. *Journal of Social Economics, 24*(1), 105–127.

Jordan, J. R., Hungerford, H. R., & Tomera, A. N. (1986). Effects of two residential environmental workshops on high school students. *Journal of Environmental Education, 18*(1), 15–22.

Kagawa, F., & Selby, D. (2012). Ready for the storm: Education for disaster risk reduction and climate change adaptation and mitigation. *Journal of Education for Sustainable Development, 6*(2), 207–217.

Mannion, G., Fenwick, A., & Lynch, J. (2013). Place-responsive pedagogy: l Learning from teachers' experiences of excursions in nature. *Environmental Education Research, 19*(6), 792–809.

Massey, D. (2007). Doreen Massey on the importance of geography – Transcript Radio 4 Today programme. Gaian Economics. Retrieved 1 November 2014 from http://gaianeconomics.blogspot.co.uk/2006/12/doreen-massey-on-importance-of.html.

Moser, S. C., & Dilling, L. (2004). Making climate hot. Environment. *Science and Policy for Sustainable Development, 46*(10), 32–46.

Norgaard, K. M. (2009). *Cognitive and Behaviour Challenges in Responding to Climate Change: Background Paper to the 2010 World Development Report.* World Bank Policy Research Working Paper Series.

Oskamp, S. (1995). Resource conservation and recycling: behaviour and policy. *Journal of Social Issues, 51*(4), 157–177.

Pelletier, L. G., Tuson, K. M., Green-Demers, I., Noels, K., & Beaton, A. M. (1998). Why are you doing things for the environment? The Motivation Towards the Environment Scale (MTES). *Journal of Applied Social Psychology, 28*(5), 437–468.

Reay, D. (2005). *Climate Change Begins at Home: Life on the Two-way Street of Global Warming.* London and New York: Macmillan.

Schunk, D. H., Meece, J. R., & Pintrich, P. R. (2012). *Motivation in Education: Theory, Research, and Applications.* London: Pearson Higher Ed.

Stern (2000). Toward a coherent theory of environmentally significant behavior. *Journal of Social Issues, 56*(3), 407–424.

United Nations Educational, Scientific and Cultural Organization (UNESCO). (2016). Education: Curriculum. Retrieved 10 February 2017 from, www.unesco.org/new/en/education/themes/strengthening-education-systems/quality-framework/core-resources/curriculum/.

Index